D0106509

Ingenious Mathematical Problems and Methods

By

L. A. Graham

Engineering Executive and Founder,
Graham Transmissions, Inc.

Dover Publications, Inc., New York

Ingenious Mathematical Problems and Methods is a new work published for the first time in 1959 by Dover Publications, Inc. The publisher is grateful to Oxford University Press for permission to reproduce illustrations from *The Oxford Nursery Rhyme Book* by Iona and Peter Opie, copyright © 1955 by Oxford University Press, Inc.

Standard Book Number: 486-20545-2
Library of Congress Catalog Card Number: 59-11841

Manufactured in the United States of America
Dover Publications, Inc.
180 Varick Street
New York, N. Y. 10014

CONTENTS

75 PROVOCATIVE PROBLEMS

25 CHALLENGING QUICKIES

MATHEMATICAL NURSERY RHYMES

Mathematical nursery rhymes are interspersed throughout PROBLEMS 1-75.

FOREWORD

THE INGENIOUS MATHEMATICAL methods and problems in this book represent the original contribution of scores of skilled mathematicians to a "Private Corner for Mathematicians," conducted by the writer for the past eighteen years in the Graham DIAL, which circulates to over 25,000 engineers and production executives. The interest shown in these methods as applied to the novel and ingenious problems appearing in this corner has been great enough to warrant republication of much of the material in more complete form in this book.

For the most part the problems were contributed by readers and it is believed that about three-fourths of them are original. Nearly all were subject to a variety of solutions, the aim being to find the method of solution which was most original, shortest, or simplest — as the case may be, and so point the path to greatest pleasure, profit and prowess in the use of mathematics, as exemplified in these problems.

Because the cardinal policy has been to avoid the trite and the ordinary and to assure at all times authentic application, it has been possible to attract a readership for our column more extensive, we believe, than that enjoyed by any other mathematical column appearing here or abroad. The most striking thing about the answers received has been their lack of similarity. In nearly every case the best solutions displayed a refreshing originality and many were superior to the one offered by the author of the problem.

In analyzing the answers received, it seems that a vast group of people, when they put aside formal mathematics at graduation, retain nevertheless a latent love and interest in the subject which are quickened and aroused by provocative problems of the sort we were able to offer. To best recapture this yen for things mathematical, we were careful to select out of hundreds of problems offered us those that not only had a new and unusual twist, calling for ingenious solution, but which involved fascinating phases of Ars Mathematica not commonly included in school curricula, such as Number Theory, Statistics, Compass Geometry, Diophantine Equations, Vectors, Networks, Inversion, and the like.

All the problems were carefully worded to avoid the abstract and were given a practical savor and narrative flavor, which further add to their appeal. Obviously it has been possible in extending the material from the confines of a "Private Corner" to the space of a book to include many interesting answers that had to be omitted from the magazine column, for lack of space. These include a scattering of wrong or roundabout answers, which when properly analyzed are often as informative to the mathematically minded as the optimum solution, as well as some refreshingly humorous contributions from those who seemed to find special interest in our columns.

Also, interspersed in these pages are a number of "Mathematical Nursery Rhymes." The first of these, opposite p. 10, appeared in "Mathematical Pie," a publication of the mathematical staff of the School for Boys, Leicester, England, and inspired the composition by DIAL readers of diverting verses of similar character, which have since appeared in each issue. Many of the droll illustrations that head the verses are taken by permission from the fascinating Oxford Nursery Rhyme Book and date back as far as two centuries. These rhymes should bring welcome diversion from the more serious matter in the volume.

75 PROVOCATIVE PROBLEMS

THE PROBLEMS THAT follow have been selected as the most interesting of those published in the Graham DIAL over a seventeen year period, judged by the number, variety and originality of solutions received. The entire group is printed in this preliminary chapter, separately from the solutions, so that the reader may select the problems that are most challenging, without having his mode of approach influenced by seeing the diagrams that illustrate the correct answers printed further on in this book.

1. THE PLUG GAUGES

This interesting shop problem, which is more difficult than appears at first glance, was originated by Edward C. Varnum, Mathematician, Barber-Colman Co., Rockford, Ill.

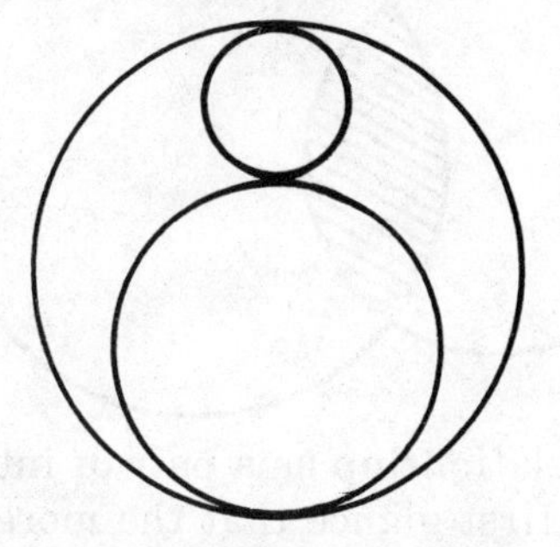

Fig. 1

A foreman noticed an inspector checking a 3″ hole with a 2″ plug and a 1″ plug and suggested that two more gauges be inserted to be sure that the fit was snug. If the new gauges are to be just alike, what should their diameters be?

2. LARGEST PRODUCT

Using the digits 1, 2, 3, 4, 5, 6, 7, 8, and 9 only once, what are the two numbers which multiply to give the largest product? To clarify, the solution might be 7463 times 98512, which uses the nine digits only once, but it isn't, because there is a pair fulfilling the conditions which gives a greater product than those two.

3. MRS. MINIVER'S PROBLEM

The following is culled from the book *Mrs. Miniver*:

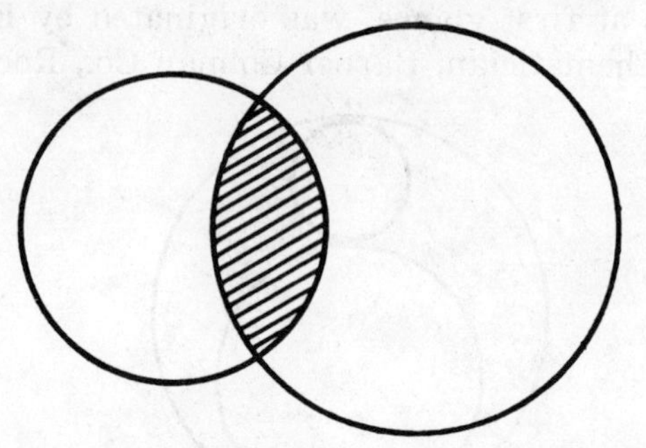

Fig. 2

"She saw every relationship as a pair of intersecting circles. It would seem at first glance that the more they overlapped the better the relationship; but this is not so. Beyond a certain point, the law of diminishing returns sets in, and there are not enough private resources left on either side to enrich the life that is shared. Probably perfection is reached when the area of the two outer crescents, added together, is exactly equal to that of the leaf-shaped piece in the middle. On paper there must be some neat mathematical formula for arriving at this; in life, none." Remembering that the two circles are rarely equal, what is the mathematical answer to Mrs. Miniver's enigma?

4. THE BRIDAL ARCH

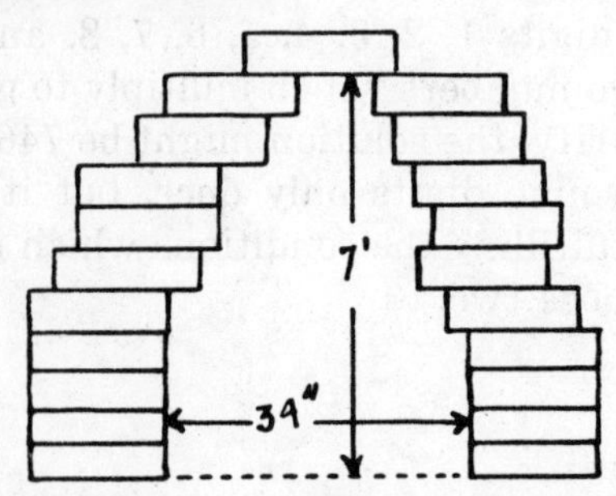

Fig. 3

An engineer whose daughter was to have a garden wedding on the morrow decided at the last minute that it would be nice to have an archway under which the bridal procession could file before the knot was tied. The opening had to be 7 feet high and at least 34 inches wide at the bottom. There was a conveniently level concrete walkway of adequate width on which to build the arch but unfortunately he found that he had just 86 bricks on hand and no mortar. The bricks, however, were in perfect condition, being very smooth and having sharp corners. Moreover, their dimensions were quite uniform, 2 x 4 x 8 inches. These favorable conditions made him decide to attempt the arch, but it seemed that no matter how he tried to arrange the bricks, they would topple off one another before the arch reached the required dimensions. Finally, his young son Euclid, who had been looking on amusedly, stepped up and told his Dad how to do the job. How?

5. THE CREEPING ROOF SHINGLE

The following problem, which was adjudged by many readers as one of the most original and fascinating in our collection, is the contribution of Dr. Stanley J. Mikina, Research Engineer, Westinghouse Research Laboratories: "A

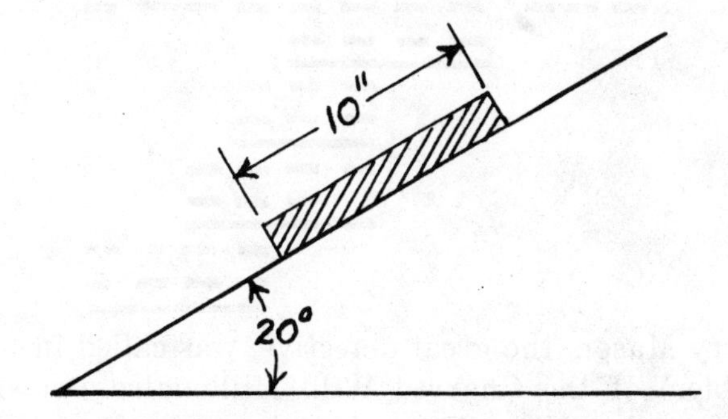

Fig. 4

flat metal shingle 10″ x 4″ x 3/16″ weighing 2 lbs. rests, with its long dimension along the slope, on a smooth wooden roof having an angle of 20 degrees to the horizontal. The shingle is not attached to the roof, but is kept from sliding down by its friction; the coefficient being .5. In the morning of a winter day the shingle is at a temperature of 0°F.

During the day it is warmed by the sun to 50°F, and at nightfall it is again chilled to 0°F. After such a cycle of heating and cooling, will the shingle have moved from its morning position by the end of the day, and if so, how much and in what direction? Assume the shingle has a thermal coefficient of expansion of $6\ (10)^{-6}$ inches per inch per degree and no expansion of the roof."

6. THE SQUARE ROOT EXTRACTOR

How could you quickly find the square root of any number without log table or slide rule or any mathematical calculation, in fact, with nothing on hand but a pair of compasses and a graduated rule? After getting the answer to the above, an original problem by George N. Smith of Perry, Ohio, go one step further and invent a "Square Root Extractor," based on your solution.

7. THE CHARRED WILL MYSTERY

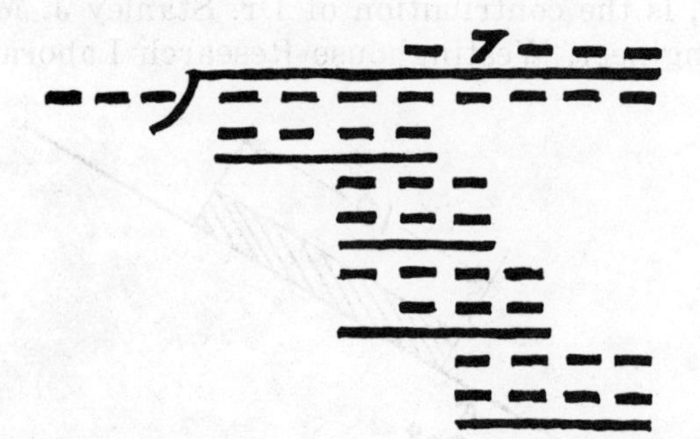

Fig. 5

Perry Mason, the great detective, was called in to solve the problem of the Charred Will. Millionaire Brown died in a Nassau fire. His will was indecipherable but a piece of charred paper was found in the old man's effects which indicated to Perry that Brown had intended to divide his estate equally among his numerous heirs. On the paper was a calculation in long division. Unfortunately only a single digit in the quotient could be identified, but the microscope showed that there had been figures in each of the spaces indicated, and that there had been no remainder. That was enough for Perry. He filled in the missing figures in the only possible way and found that they checked exactly with the old man's fortune and the number of his heirs. What was Perry's reasoning and how many heirs were there?

8. FROM POLE TO POLE

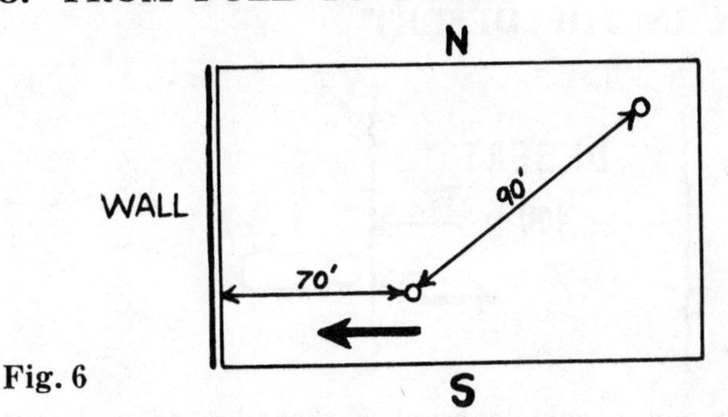

Fig. 6

The kids were in the playground playing games when the teacher, Miss Pythagoras Jones, thought up a new one. "Come here, boys", she said, "and gather 'round this pole. We are going to see who is the best runner. I will time you with my stop watch. Each boy in turn must start at this pole, run to any point in the wall, make a chalk mark and then run to that pole over there. Let's go!" The starting pole was 70 feet east of the wall which ran north and south, and the finishing pole was 90 feet due north-east of the starting pole. All the kids were equally fast but little Euclid Smith had the best time. What path did he take?

9. THE BALLOT-BOX PROBLEM

An election for village president was won handily by Smith with 300 votes to 200 for the other candidate, Jones. In tallying the votes one by one at the village hall when the polls closed, what is the chance that at least once during the count after the first vote was tallied, did Jones, though ultimately defeated, have the same recorded total as Smith? Further, what would this chance be if the winning and losing totals were a and b, rather than 300 and 200? This interesting Ballot-Box Problem, with his own novel mode of solution, was contributed by Benjamin Graham, noted authority on investment analysis.

10. TRUCK IN THE DESERT

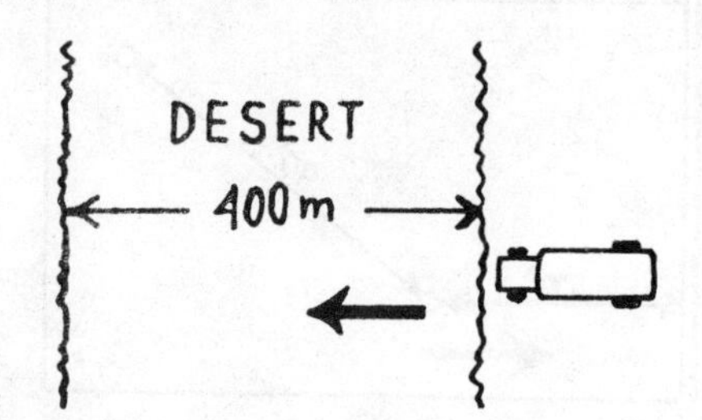

Fig. 7

A heavy motor vehicle reaches the edge of a desert 400 miles wide. The vehicle averages only one mile to the gallon of gas, and the total available gasoline tank capacity, including extra cans, is 180 gallons, so it is apparent that gasoline dumps will have to be established in the desert. There is ample gas to be had at the desert edge. With wise planning of the operation what is the least gas consumption necessary to get the vehicle across the desert?

11. THE UNRULY BISECTION

Mr. Howard G. Taylor, of Diamond Chain & Mfg. Company, found himself in an unusual dilemma. He had a straight line to bisect but no tools whatever with which to do the job but a pair of compasses. How did he do it?

MATHEMATICAL NURSERY RHYME No. 1

Fiddle de dum, fiddle de dee,
A ring round the moon is π times D;
But if a hole you want repaired,
You use the formula πr^2.

MATHEMATICAL NURSERY RHYME No. 2

Rock-a-bye baby in the tree top,
As a compound pendulum, you are a flop.
Your center of percussion is safe and low,
As one may see when the wind doth blow.
Your frequency of vibration is pretty small,
Frankly, I don't think you'll fall at all.

MATHEMATICAL NURSERY RHYME No. 3

Ride a fast plane
And don't spare the cash,
To see the math wizard do sums in a flash.
Roots on his fingers
And powers on his toes, —
He carries log tables
Wherever he goes.

MATHEMATICAL NURSERY RHYME No. 4

Little Bo-Peep has lost her sheep,
Totaling 10100;
She counted by twos, meaning fewer to lose,
But we trust that all twenty get caught.

MATHEMATICAL NURSERY RHYME No. 5

A diller, a dollar,
A witless trig scholar
On a ladder against a wall.
If length over height
Gives an angle too slight,
The cosecant may prove his downfall.

MATHEMATICAL NURSERY RHYME No. 6

Where are you going to, my pretty maid?
I'm going a-milking, sir, she said.
How many gallons to drink when you're done?
Divide cubic inches by 231.

12. START OF THE SNOW

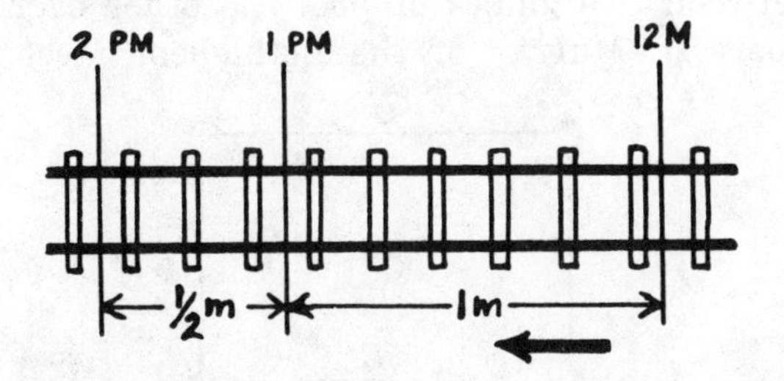

Fig. 8

Snow starts to fall in a certain city in the forenoon and falls at a constant rate all day. At noon a snow plow starts to clear a streetcar track. The plow goes a mile during the first hour and a half mile during the second hour. What time did it start to snow?

13. THE ROOKIE ELECTRICIAN

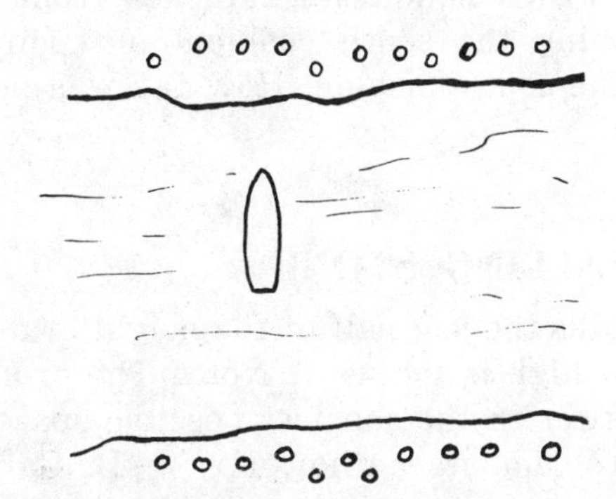

Fig. 9

A rookie electrician was ordered by his foreman to row across a stream with a bundle of 21 wires and connect them to a control board on the other side. The poor novice got back to his boss and admitted shamefacedly that he hadn't labeled the wires. "Get going," shouted the foreman "and label those wires by yourself with the least rowing and use no unnecessary instruments, or you are fired." How did he do it?

14. THE POOL TABLE

"Between two games of pool the other evening," said Mr. Robert H. Murray of the Bethlehem Steel Company,

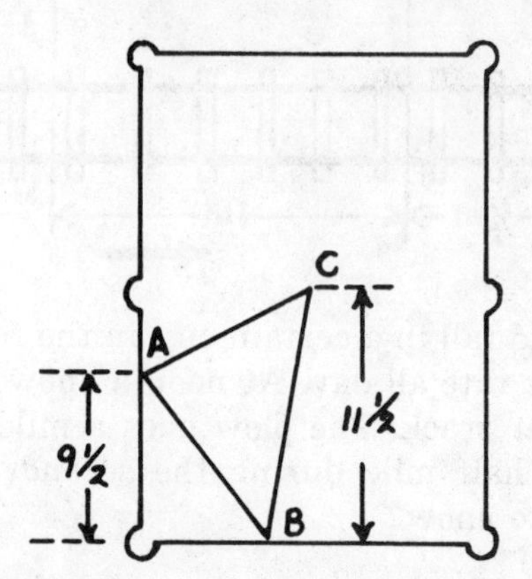

Fig. 10

"I noticed that the triangle (call it ABC and forget its rounded corners) was in a corner of the table with point A touching the west cushion 9-1/2 inches from the pocket; point B touching the south cushion and point C 11-1/2 inches from the south cushion. How far was point B from the pocket?"

15. THE FALLING STONE

A stone falls the last half of the height of a wall in half a second. How high is the wall? Note: The problem here is to get the answer in the shortest possible space. In contributing the problem, its author, Mr. J. B. Gaffney of the Fuller Company, wrote: "It is simple — merely requiring high school algebra and Newton's law of falling bodies. Just the same it is not so simple to do it the easy way. Why not specify that the complete solution be submitted on not over one-half of the back of an ordinary business card? It can be done, including sketch and complete calculation, on the back of an air mail stamp. The complete calculation on one-half of a business card is enclosed herewith."

16. MAGIC MULTIPLIERS

Wrote Mr. Erwin Bunzel, Chief Draftsman, Standard Brands, Inc.; "What is the shortest time in which the following multiplication can be accomplished? 4109589041096 x 83. It can be done in one second, by just placing the 3 (of 83) in front and the 8 (of 83) at the end of the multiplicand, for the answer is 341095890410968. What is the explanation, and are there any more cases where the product obtained by multiplying a number by a two-digit multiplier can be had in this lightning fashion?"

17. THE FOUR FOURS

Said Major Hitch of the British Army, "Write down an expression using only the figure 4 four times, which will equal 71. Any mathematical symbols whatever can be used in connection with the figures — such as roots, decimals, factorials, fractions, etc. but no other figure may be used except 4 and this is used 4 times to give the required answer of 71."

18. PROBLEM OF THE EASIEST THROW

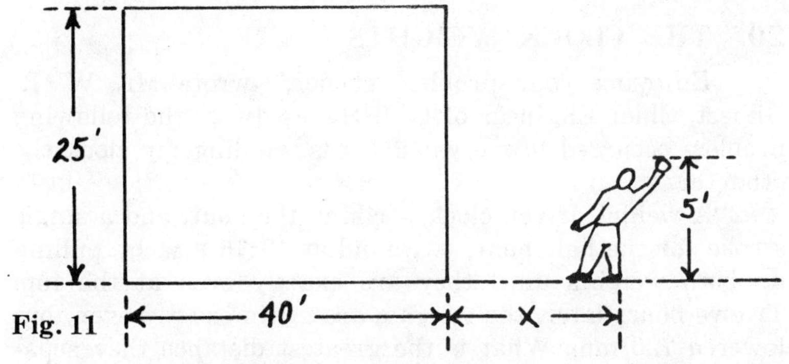

Fig. 11

It will remembered (see Problem 8) that Little Euclid had won a historic race in the schoolyard. We must further relate that the little flat-topped red schoolhouse stood 25 ft. high and 40 ft. wide and Euclid then felt constrained to celebrate his victory by throwing a ball clear over the roof with the least effort. How far away from the wall did he take his stand, with ball in hand, outstretched 5 feet above terra firma?

19. THE SLICED TRIANGLE

(1) Divide a triangle into three equal areas with two lines drawn parallel to the base (using compasses and straight edge only). (2) Then as a second problem divide a triangle into five equal areas with four lines drawn parallel to the base. And (3) applying the method used in part 2, solve the problem of old Agricola's inheritance. Agricola died, leaving his property to his four sons, aged 20, 30, 40 and 50 respectively. The property had a frontage of 800 yds., a depth of 400 yds. at one end, and 800 yds. at the other. The ends were parallel and at right angles to the frontage line. The plot was to be divided by three lines parallel to the edges into areas proportional to the sons' ages. How was the postion of the lines found on paper, geometrically, using compass and rule?

20. THE CLOCK WEIGHTS

"Enjoying your problem corner," wrote Mr. W. R. Gilbert, Chief Engineer of C. I. Hayes, Inc., "the following problem occurred to me when I was winding my clock the other night:

"A weight-driven clock, striking the hour, and a single stroke for the half hour, is wound at 10:15 P.M. by pulling up both weights until they are exactly even at the top. Twelve hours later, the weights are again exactly even, but lowered 720 mm. What is the greatest distance they separated during the interval?" (Note: This problem is readily solved progressively by cut-and-try, using a chart or table, but it is also subject to analytic handling by the calculus.)

21. THE MISSING BEECH

(Contributed by Mr. Yondin, Amperex Electronic Products)

Fig. 12

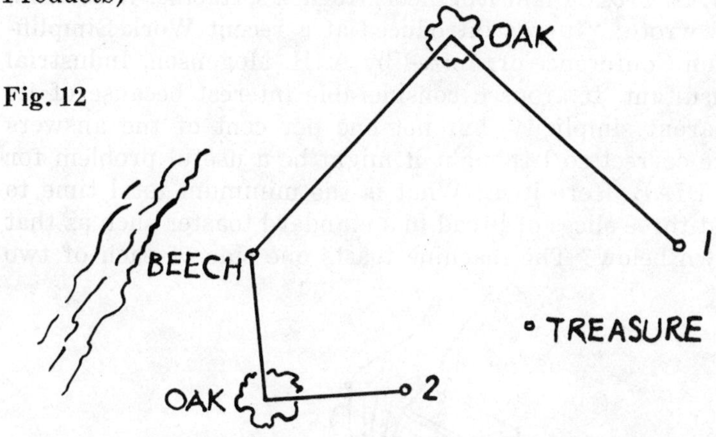

Captain Flint had to hurriedly secrete a treasure on a lonely island. No time for maps. There were three trees forming a triangle; a beech nearest the water's edge and an oak on either side. Said the captain to Long John: "Stretch a rope from the beech to this first oak and then pace a distance inland from the oak at right angles to the rope, and equal to the rope's length. That will be our point No. 1. You, Silver, do the same at the other oak. Stretch a rope from beech to oak and walk a distance inland from the oak at right angles to the rope and equal to its length. Good. That's our point No. 2. Now, boys, we will bury the treasure halfway between our two points. Okay, cut away the ropes and off to the boats. Let's go."

Six months later, Long John, accompanied by the cabin boy, stole back to the island to double-cross the gang and swipe the treasure. He knew the scheme but, sad to relate, the beech that had been the reference point had been completely washed away in a storm and only the oaks remained. Long John was crestfallen. But by great good luck the cabin boy was none other than our Euclid Smith. "Don't worry, John", cried Euclid. "I'll find you the treasure without the beech." And, he did — in a jiffy. How?

22. THE TOASTER PROBLEM

This is an interesting time study problem contributed by J. H. Cross, Plant Engineer, Horlick's, Racine, Wisconsin, who wrote: "It was introduced at a recent Work Simplification Conference organized by A. H. Mogensen, Industrial Consultant. It aroused considerable interest because of its apparent simplicity, but not one per cent of the answers were correct so I thought it might be a useful problem for the DIAL. Here it is: What is the minimum total time to toast three slices of bread in a standard toaster such as that shown below? The machine toasts one side of each of two

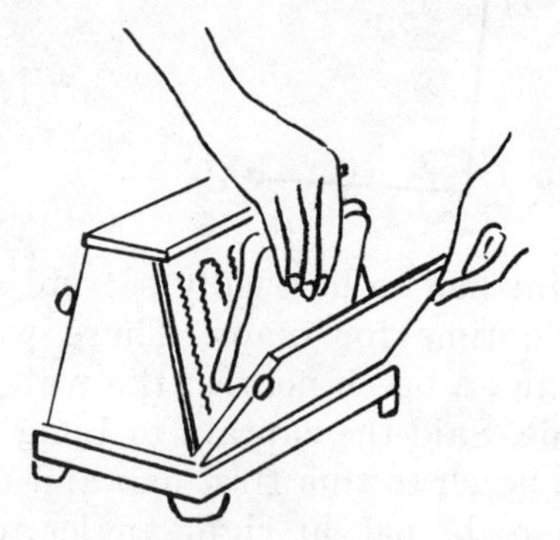

Fig. 13

pieces at the same time. It takes two hands to insert or remove each slice, as illustrated. To turn the slice over it is merely necessary to push the toaster door all the way down, and allow the spring to bring it back. Thus two slices can be turned at the same time, but only one can be inserted or removed. The time to toast a side is exactly .50 minutes. Time to turn over is .02 minutes. Time to remove toasted slice and place on plate is .05 minutes, and the time to secure a piece of bread and place in the toaster is .05 minutes. The problem is to find the shortest possible time required to toast three slices on both sides, starting with bread on plate, and returning toast to plate. Assume toaster is warmed and ready to go."

23. THE COMPLAINING GOAT

This problem was originated by R. A. Harrington, B. F. Goodrich Research Center, Brecksville, Ohio. It seems

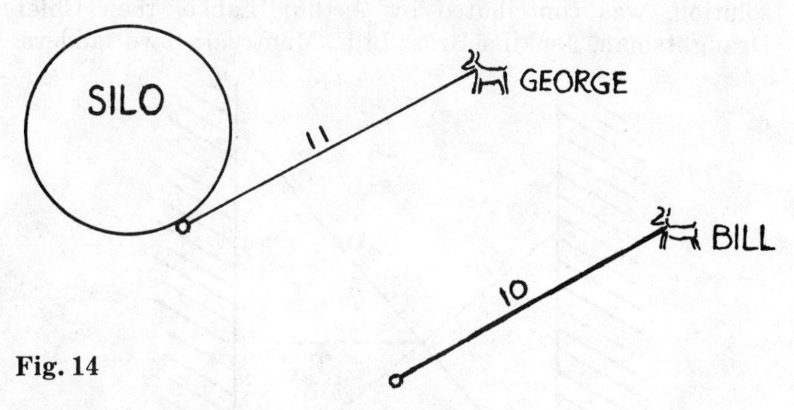

Fig. 14

that George and another goat, Bill, were both tethered in a pasture, and fell to arguing about their relative grazing areas. "It is true," bemoaned George," that my rope is ten per cent longer than yours, but mine is tied to a ring on the outside wall of this circular silo and I can just reach the point on the wall diagonally opposite the ring, so I get no grass at all in that particular direction. You on the other hand are tethered to a stake in the middle of the field and can graze over a complete circle, so it seems you are much better off than I." The problem is to explain to George just how his grazing area compares with Bill's.

24. THE TIMED SIGNALS

This innocent-looking puzzler for automobile drivers was originated by D. D. Wall, Los Angeles and requires a much more complete analysis for the full solution than appears at first blush: "If traffic signals are 'set for 30,' at what other constant speeds may one travel? Assume that the signals are evenly spaced, all change at the same time, and that they alternate in color spacewise as well as timewise."

25. THE TWO LADDERS

This interesting variation of the familiar two-ladder problem, which requires ingenuity to devise even the simplest solution, was contributed by Arthur Labbe, then Chief Draughtsman, Jenkins Bros. Ltd., Montreal: Two ladders,

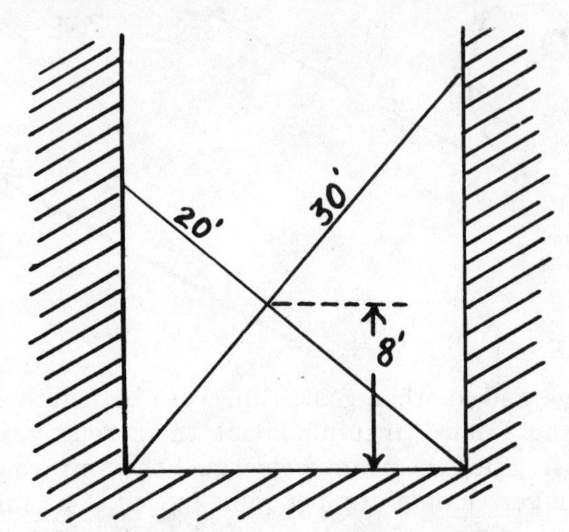

Fig. 15

20 ft. and 30 ft., cross at a point 8 ft, above a lane. Each reaches from the base of one wall to some point on the opposite wall. How wide is the lane?

26. THE HATCHECK GIRL

This interesting problem, originated in the form used in the DIAL by W. Weston Meyer, Research Engineer, Bundy Tubing Co., Birmingham, Michigan, not only has the fascination of leading to a concrete result from apparently insufficient data but brings out one of the most interesting relations in all number. It tells of a hat-check girl in a large theater who mixed up all her checks and proceeded to hand out the hats at random at the end of the show. If the customers consented to this indignity, what is the chance that not a single one would go home with his own headpiece? Incidentally, Mr. Meyer asked our readers to check their answer by taking two packs of cards, shuffling them separately, then turning over one card at a time in each until the cards match in each pack. If you are willing to do this for a dozen or perhaps two dozen times, and after completing your solution, record the number of times out of your total trials when no matching occurred throughout the two packs; you will be able to record for posterity how closely nature conforms to the postulates of probability.

27. GEARS IN THE JUNGLE

A. Hollander of Byron-Jackson Co., Los Angeles, relates: "During the past war, an American company of engineers found themselves in the Australian jungle faced with the dire necessity of building a 6 to 1 reduction gear. They looked around for gears and found a dozen, strong enough for even half their width — the only trouble was that they were all alike. The captain cursed; the sergeant stepped up and sketched the way to do it. What was the sergeant's plan?" This prøblem requires no elaborate "gearology" or engineering but does call for some imagination and originality.

28. EUCLID'S PUT-PUT PROBLEM

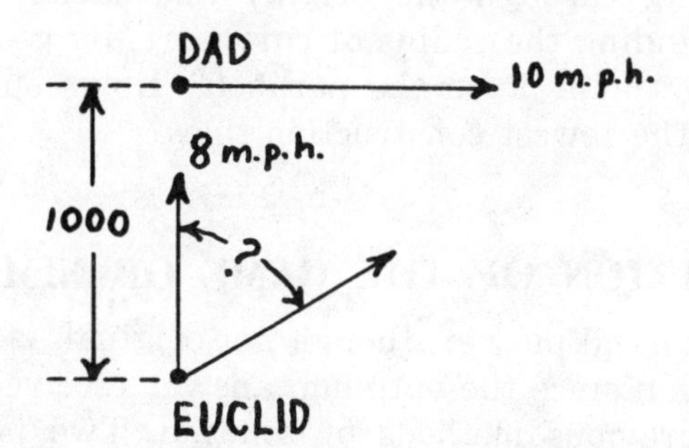

Fig. 16

Little Euclid was recently running his put-put at its top speed of 8 miles per hour when he saw his dad in the launch going 10 miles per hour, 1000 yards dead ahead of him, and crossing his path at right angles. It was quite important that Euclid get within the closest possible hailing distance, to attract Dad's attention and signal that he wouldn't be home for supper but would stay out fishing. Euclid immediately altered his course by the required angle. What was the angle and how close did he get to Pop's boat?

29. THE DRAFTSMAN'S PARABOLA

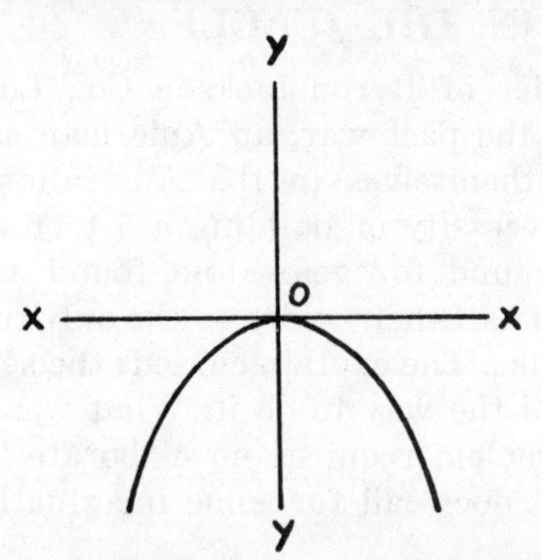

Fig. 17

This tells of the quandary of the draftsman who found himself with a drawing of a parabola (for convenience, let it be below the X axis, symmetrical with respect to the Y axis and going through the origin) and faced with the necessity of finding the radius of curvature, by geometrical means, at any point P on the parabola. How could he do the job with the fewest construction lines?

30. VARIATION OF THE GAME OF NIM

This historical puzzler, though not original, is included in our collection since the optimum answer received proved superior to previous methods of solution. Two players A and B have three separate piles of coins on a table before them, each pile having a random number of coins. Each player in turn removes one or more coins from *one only* of the piles (he can remove one complete pile, if he wants to) with the object of finally forcing his opponent to remove the last single coin. How can you play this game to make certain of winning?

31. EQUIDISTANT MEETING POINTS

Francis L. Miksa, Aurora, Illinois, contributes the following problem: A rancher bought a triangular piece of

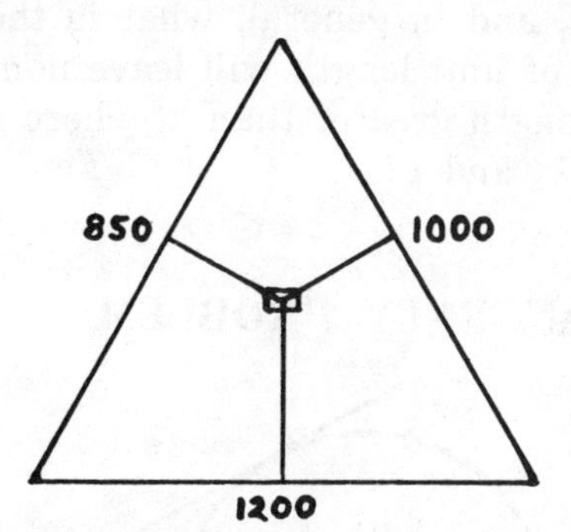

Fig. 18

land whose sides were 1200, 1000 and 850 rods. After building his house, he laid out a drive from it to the nearest point on each of the roads that bordered his land and was greatly surprised to find that the three meeting points were exactly the same distance from each other. What was that distance?

32. CUTS IN A LENGTH OF WIRE

An extremely interesting problem about Little Euclid and a triangular fence plot was originated by Theodore R. Goodman, Buffalo: It appears that Euclid's dad wanted an outdoor playpen for the little ones and thought it would be simplest to cut the roll of fence he already had into three lengths and fence off a triangular plot. He was about ready to take his two sets of snips with the wirecutter when little Euclid, who was watching the procedure, blandly remarked,

"How do you know your three lengths will make a triangle?"

"Oh, I'll take a chance," said Dad.

If his two cuts were made entirely at random, just what chance was Dad taking?

When this problem appeared in the DIAL it attracted an unusually great variety of solutions, including one by Samuel M. Sherman, Philadelphia which led to his devising a supplementary problem of a more general nature, which may be solved as a corollary to the one above. What is the chance that three random cuts in a length of wire will form a quadrilateral, and in general, what is the chance that n cuts in a piece of unit length will leave none of the (n+1) pieces with a length greater than x, where x may have any value between ½ and 1?

33. THE TANGENCY PROBLEM

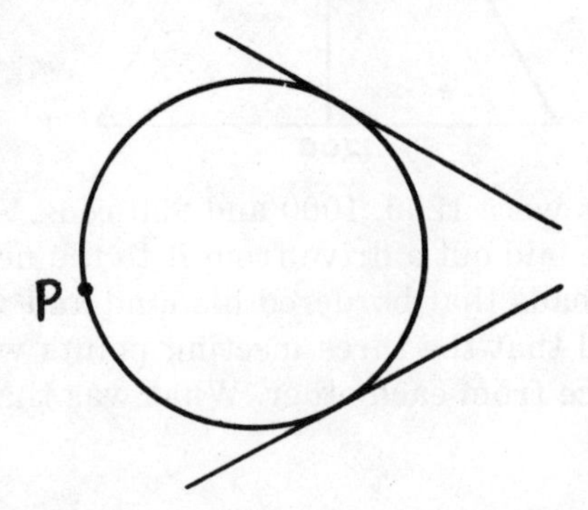

Fig. 19

This apparently simple problem in plane geometry uncovered a unique mode of solution of considerably more interest than the conventional approach: Pass a circle through a given point and tangent to two intersecting straight lines (and try to do it without use of the usual "helper" circle).

MATHEMATICAL NURSERY RHYME No. 7

Sing a song of sixpence—
A mathman full of rye.
Four times twenty square feet
Multiplied by π
Gives the total ground he covers
While weaving an ellipse;
His path would have no area,
If he had no nips.

MATHEMATICAL NURSERY RHYME No. 8

Hey diddle, diddle,
The cat and the fiddle,
The cow jumped into the blue;
Her leap into action
Took plenty of traction,
The product of Force times mew.

MATHEMATICAL NURSERY RHYME No. 9

Little Jack Horner sat in a corner,
Trying to evaluate π.
He disdained rule of thumb,
Found an infinite sum,
And exclaimed "It's REAL, nary an i."

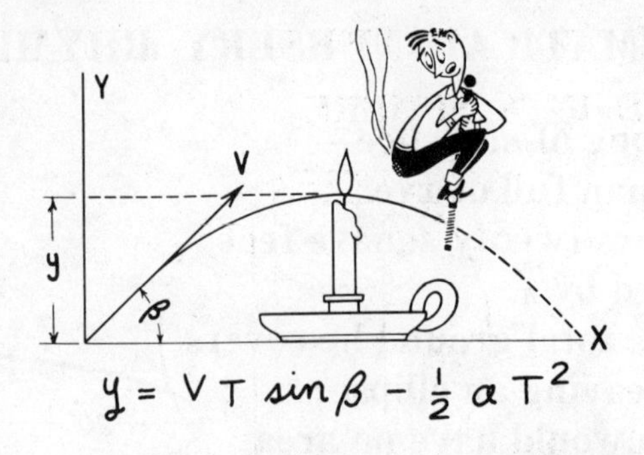

MATHEMATICAL NURSERY RHYME No. 10

Jack be nimble, Jack be quick,
Jack jump over the candlestick.
But figure out β and also time T,
"a" due to gravity, velocity V,
And don't forget $y = VT \sin \beta$
Minus $\frac{1}{2} aT^2$, or you'll regret later.
Figure trajectory right to the inch
Or it might be a "singe" instead of a cinch!

MATHEMATICAL NURSERY RHYME No. 11

RUB-A-DUB-DUB,
Three men in a tub,
Useful volume of tub must be
Weight of tub plus the fellows
(If you disregard billows)
Over specific weight of the sea.

MATHEMATICAL NURSERY RHYME No. 12

Little Miss Muffet
Sat on a tuffet,
Counting her surds, and say,
Along came a binar
And counted beside her,
Which frightened Miss Muffet away.

34. HOLE IN A SPHERE

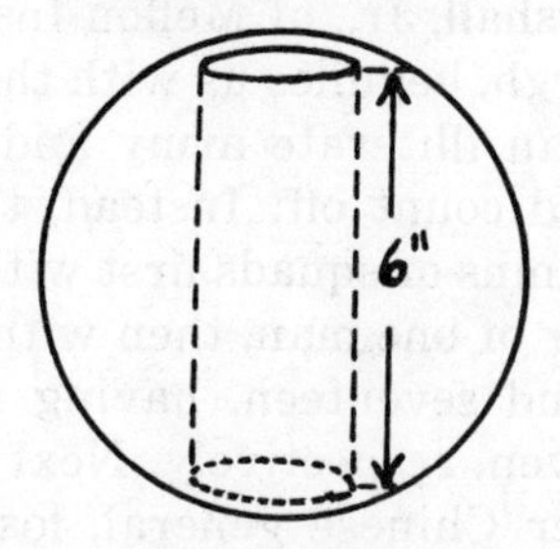

Fig. 20

This problem should particularly intrigue readers that have anything to do with machine tools:—"A hole six inches long is drilled clear through the center of a solid sphere. What is the volume of material remaining?" After getting your answer (and we haven't left anything out) try to give a physical explanation and/or interpretation of the surprising result.

35. THE FARM INHERITANCE

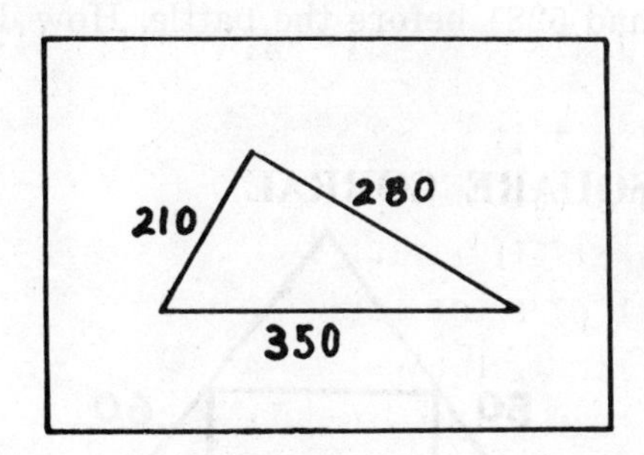

Fig. 21

H. M. Edmunds, of R. Hoe & Co., New York, tells of a farmer who bequeathed to his youngest son Euclid a section of his property in the form of a right triangle, with sides equal to 280, 210 and 350 yards, and with the proviso that the boy could take an additional section of land in the form of another right triangle of equal area whose dimensions were different but also integers, provided he could compute its dimensions, which he did. How?

36. THE CHINESE GENERAL

Byron O. Marshall, Jr., of Mellon Institute of Industrial Research, Pittsburgh, beguiles us with the tale of "a Chinese general who had an illiterate army and couldn't order his men to line up and count off. Instead, the general had his men line up in columns of squads first with four men abreast, giving a remainder of one man, then with five abreast, seven abreast, eleven, and seventeen, having remainders of one, three, one and eleven, respectively. Next week he got into a scrap with another Chinese general, losing some men and taking some prisoners. The tally is:

Column of squads	Remainder before Battle	Remainder after Battle
4	1	3
5	1	1
7	3	2
11	1	2
17	11	12

Fig. 22

The general had 5281 before the battle. How did he fare?"

37. THE SQUARE CORRAL

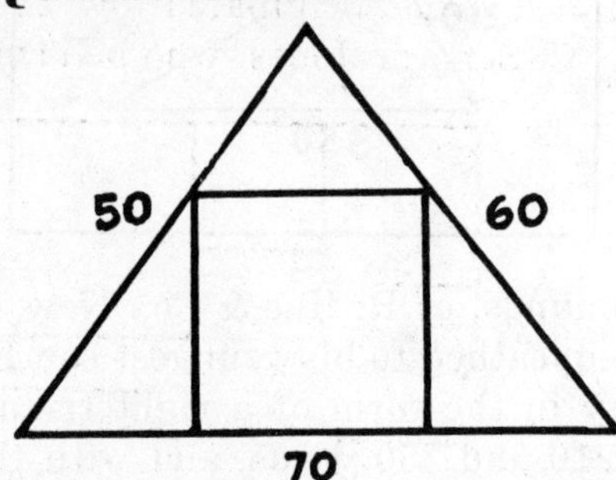

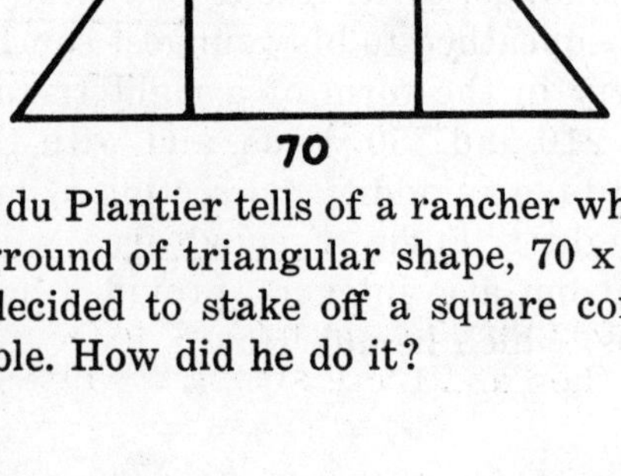

Fig. 23

Mr. D. A. du Plantier tells of a rancher who had an outlying plot of ground of triangular shape, 70 x 60 x 50 rods, on which he decided to stake off a square corral, to be as large as possible. How did he do it?

38. CAPTIVATING PROBLEM IN NAVIGATION

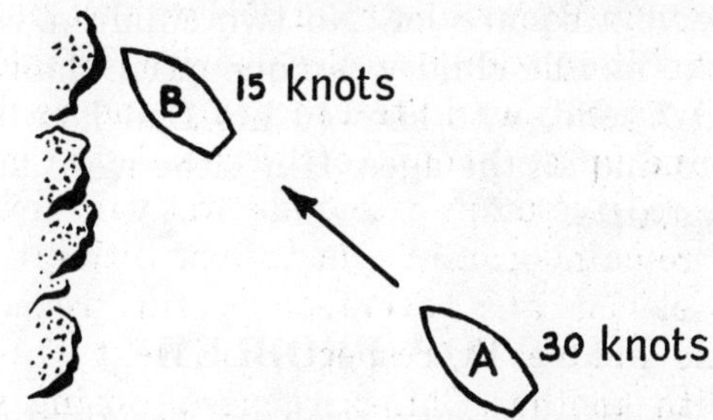

Fig. 24

Ship A is chasing ship B and making 30 knots to 15 for the pursued. Neither is equipped with radar. Ship B enters a cloud bank which makes its further visibility from ship A impossible. Captain A correctly assumes that Captain B will take advantage of the fog to immediately change his course and will maintain his new direction unchanged at full speed. Based on this assumption, what plan should Captain A follow to insure that he will intercept ship B?

39. THE SQUARE-DEAL SENATOR

This interesting problem in networks, which may be a bit too advanced for some readers of this book, but nevertheless contributes a useful method of analysis for the mathematically-minded, was originated by F. K. Bowers of Vancouver, B. C. Senator Jones, who believed in the Square

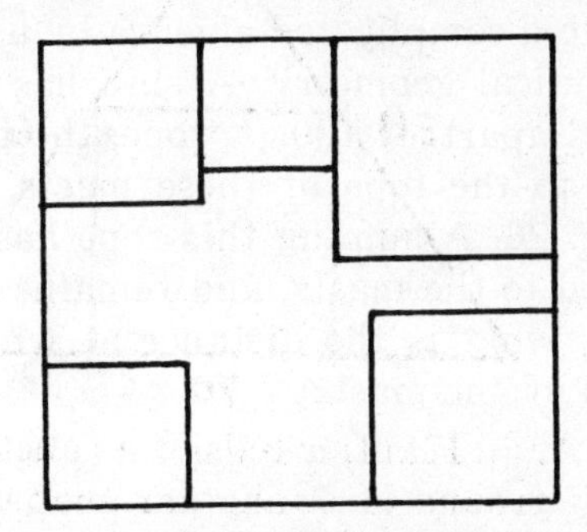

Fig. 25

Deal, had a rectangular piece of property which he decided to divide up then and there among his nine children in proportion to their ages (nearest integer), provided each child could receive a square lot. No two children were the same age and the middle child was none other than our little unwhiskered Euclid, who showed his Dad how to do it. It is required to find all the ages. Hint: the word "unwhiskered" is of importance.

40. THE MAST-TOP PROBLEM

This relatively easy problem, originated by H. F. Spillner, Chief Engineer, Johnston & Jennings Co., Cleve-

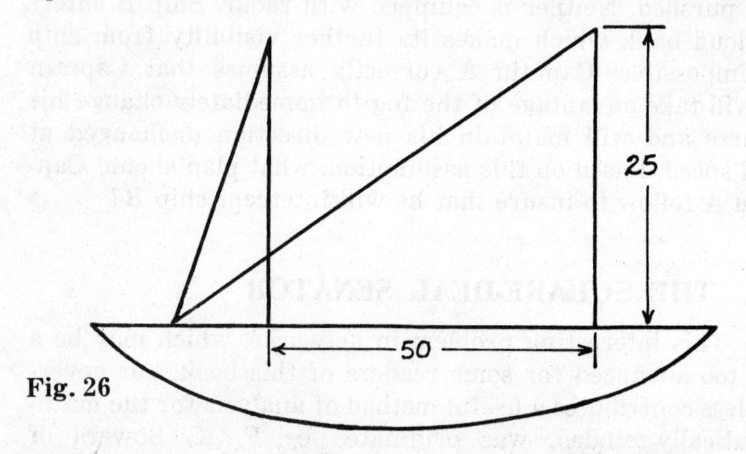

Fig. 26

land, is especially interesting as an example of how it is possible to avoid complicated algebra by using familiar relations in analytical geometry. A ship has two masts, 25 ft. tall and 50 ft. apart. A 100 ft. rope, the two ends of which are fastened to the tops of these masts, is pulled taut as shown in Fig. 26. Assuming this rope has lost no length in being fastened to the masts, and remains in the same plane as the masts, what is the distance at which it touches the deck forward of one mast?

41. THE CHANGED BASE

Mr. A. S. Hendler, Troy, N. Y.,originated this problem about logarithms: Most logarithmic problems in engineering use the base 10, as in slide rule computations. Mr. Hendler observed that the logarithm of any positive number to that base is always less than the number, and wondered if the base were some positive number other than 10, how big it would have to be for this condition to still exist.

42. MATHMAN'S GREETING CARD

This problem in simple addition is contributed in the form of a unique greeting card by Dick and Josephine Andree of the University of Oklahoma: "Given that E^2 equals H and that each symbol represents a different digit, find (and prove that it is the only solution) the key to the following addition problem:

```
      D & J
  A N D R E E
      S E N D
  -----------
    C H E E R
```

Fig. 27

43. MEETING OF THE PLOWS

This problem, originated by M. S. Klamkin, Brooklyn, is similar to but considerably more difficult than our No. 12, "Start of the Snow": It had started snowing before noon and three plows set out at noon, 1 o'clock, and 2 o'clock, respectively, along the same path. If at some later time they all came together simultaneously, find the time of meeting and also the time it started snowing." The solution is based on the natural assumption that it snows at a constant rate and that each plow removes snow at the same constant rate.

44. THE DIVIDED CHECKERBOARD

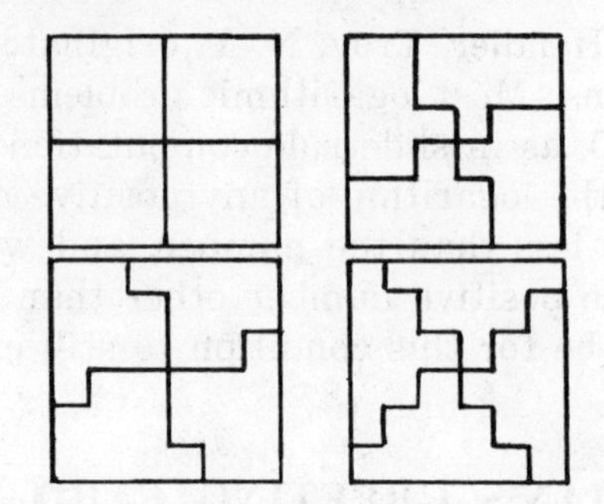

Fig. 28

Although this problem is strictly not one in mathematics, it is surprising to find how well it responds in varied ways to mathematical analysis and careful geometric layout. Originated by a teacher of mathematics, Howard D. Grossman, of New York, it reads as follows: In how many ways can a 6x 6 checkerboard be cut up along the lines of the board into four identical pieces? The four pieces must be the same when placed on top of one another — after cutting — without turning any of them over (a few typical cuttings are shown in Fig. 28).

45. TWENTY QUESTIONS

This problem, originated by H. D. Larsen, of Albion College, is an interesting variation of the popular game of twenty questions. Mr. Larsen says, "Suppose I think of a number which you are to determine by asking me not more than twenty questions, each of which can be answered by only 'yes' or 'no'. What is the largest number I should be permitted to choose so that you may determine it in twenty questions?"

46. CHILDREN AT PLAY

This narrative exercise in logic, a type that involves integral solutions expertly arrived at, was contributed by L. R. Ford, Illinois Institute of Technology:

"Are those your children I hear playing in the garden?" asked the visitor.

"There are really four families of children," replied the host. "Mine is the largest, my brother's family is smaller, my sister's is smaller still, and my cousin's is the smallest of all. They are playing drop the handkerchief." He went on, "They prefer baseball but there are not enough children to make two teams. Curiously enough," he mused, "the product of the numbers in the four groups is my house number, which you saw when you came in."

"I am something of a mathematician," said the visitor, "let me see whether I can find the number of children in the various families." After figuring for a time he said, "I need more information. Does your cousin's family consist of a single child?" The host answered his question, whereupon the visitor said, "Knowing your house number and knowing the answer to my question, I can now deduce the exact number of children in each family."

How many children were there in each of the four families?

47. THE FREE PLOT

This unique problem, originated by Royes Salmon, of Fullerton, Cal., led to more jumping to wrong conclusions than most of our puzzlers. Rancher Smith, who had a square piece of land, a mile on each side, offered to sell all or any part of it, charging nothing for the land, provided the buyer would let Smith fence the part he bought at $1.00 per ft. Farmer Brown was interested in the proposition, but wanted to be sure that he paid the lowest possible price per acre. What plot did he take?

48. THE BALANCED PLANK

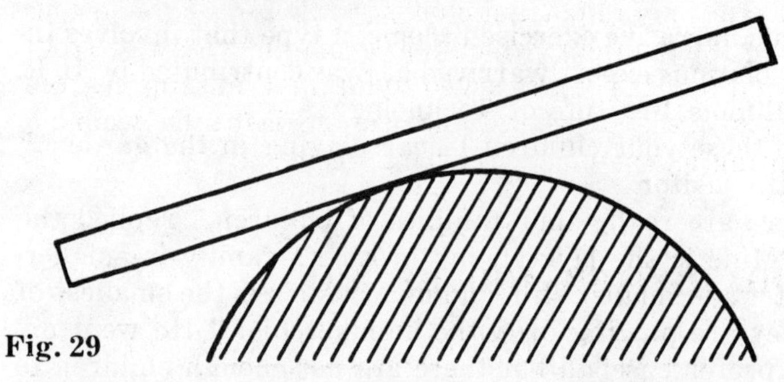

Fig. 29

A horizontal plank (see Fig. 29) is balanced on a cylindrical surface whose elements are perpendicular to the length of the plank. What is the cross section of the cylindrical surface if the plank has neutral equilibrium, and when will the plank start to slip? It will be remembered that in "neutral" equilibrium, the body remains in equilibrium after displacement; in constrast to "stable," where on being slightly displaced it tends to return to its original position; or "unstable," where it tends to move farther from that position.

49. THE RECURRING DECIMAL

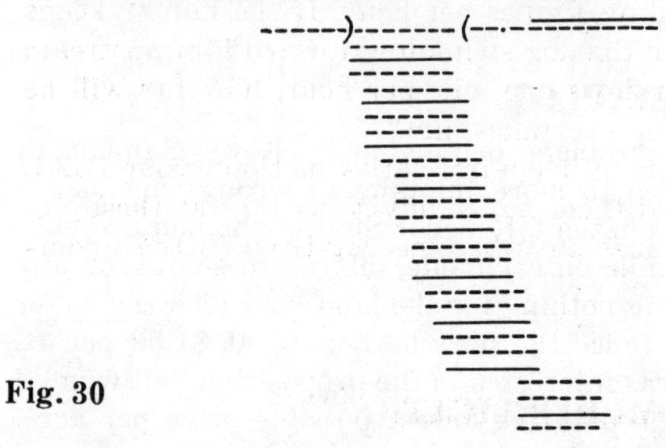

Fig. 30

This problem, contributed by Norman Wickstrand, is similar to No. 7, "The Charred Will Mystery," but goes it one better by omitting all digits in the divisor, dividend, and quotient. The only important clue is the nine-place recurring decimal in the quotient, which starts one figure away from the decimal point. You are asked to fill in all missing figures (see Fig. 30) ; the horizontal line over the last nine decimals representing a nine-place recurring decimal.

50. THE HUNTER AND HIS DOG

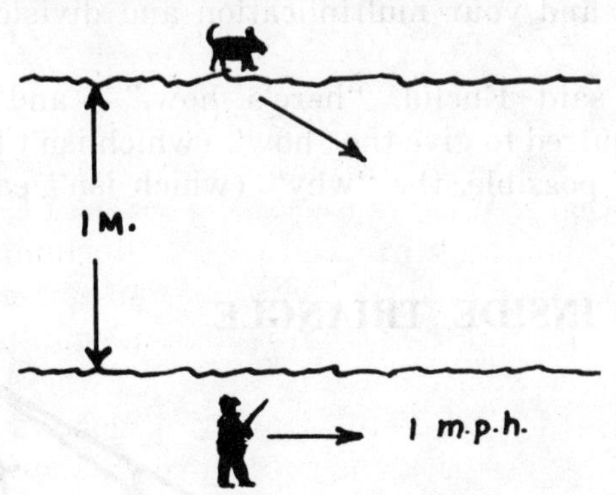

Fig. 31

A hunter was walking up one side of a mile-wide strean at one mile per hour when his dog appeared directly opposite him on the other side, jumped in, and swam toward him at a still water speed of 3 miles per hour. If the hunter keeps on walking — with the dog swimming toward him upstream — and the stream flows one mile per hour, how far will he walk before the dog overtakes him?

In offering this problem in relative motion to our DIAL readers we warned them not to advise us that at these figures the animal must have become dog-tired in the undertaking or overtaking.

51. RUSSIAN MULTIPLICATION

This problem is another example of the magic of the number 2, which formed the theme of the Nim problem, No.

30, and the problem of the Twenty Questions, No. 45. It appears that the little children were having their arithmetic lesson. They had learned addition and subtraction and had just been taught how to multiply by 2 and how to divide by 2, but that was as far as they had gone. Up rose little Euclid and said, "Now I can multiply any two numbers at all, almost as fast as my older brother Pythagoras can."

"That's silly," said the teacher. "How can you quickly multiply 85 x 76, for example, if you only know addition and subtraction and your multiplication and division tables up to 2."

"Okay," said Euclid, "here's how," — and he did it. You are required to give the "how" (which isn't too difficult) and also, if possible, the "why" (which isn't easy).

52. THE INSIDE TRIANGLE

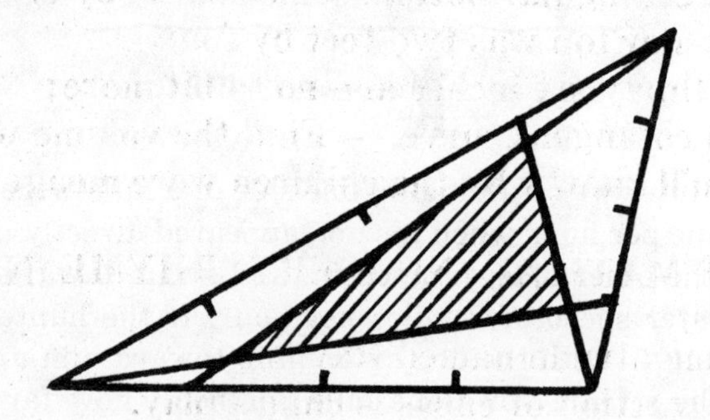

Fig. 32

This interesting geometrical problem was originated by B. L. Burton, of Los Alamos, New Mexico: The three sides of any triangle are divided into four equal parts. A line is drawn from each vertex to the ¼ position on the opposite side (see Fig. 32). This gives a triangle within the main triangle (cross-hatched area). Find the ratio of the area of this triangle to the original, and also the ratio if each side is divided into n equal parts.

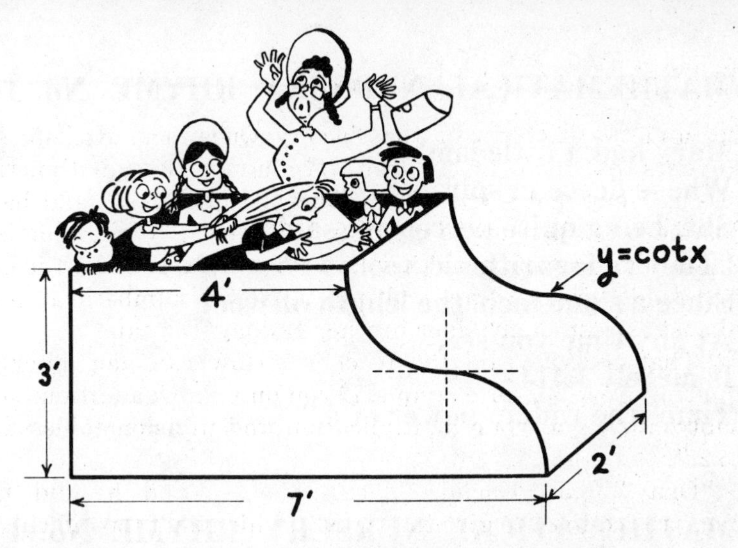

MATHEMATICAL NURSERY RHYME No. 13

There was an old woman who lived in a shoe
Whose rectangular bottom scaled seven by two;
Rectangular top was two feet by four,
Height thirty-six inches and not a bit more;
Vamp a cotangent curve. — Find the volume within
And you'll know why the children were meager and thin.

MATHEMATICAL NURSERY RHYME No. 14

A mathematician named Ray
Says extraction of cubes is child's play.
You don't need equations
Or long calculations
Just hot water to run on the tray.

MATHEMATICAL NURSERY RHYME No. 15

Humpty Dumpty sat on a wall,
Wondering how hard he would fall.
Force times time, you will agree,
Is equal to mass times velocity.

MATHEMATICAL NURSERY RHYME No. 16

Mary had a little lamb
Whose fleece in spirals grew;
She, being quite perceptive said,
"They're logarithmic, too.
Since a's one inch, the length of wool
At any time you see,
Is merely 1.414
Times the radian power of e."

MATHEMATICAL NURSERY RHYME No. 17

Jaguar Bill raced up the hill, —
his speed was 83;
At 104, intent on more,
the turn he did not see.
His drop through the air
was the root of the square
Of the length of his climb
to the edge
Less the square of the base,
from the start of his race
To the point
where the crew had to dredge.

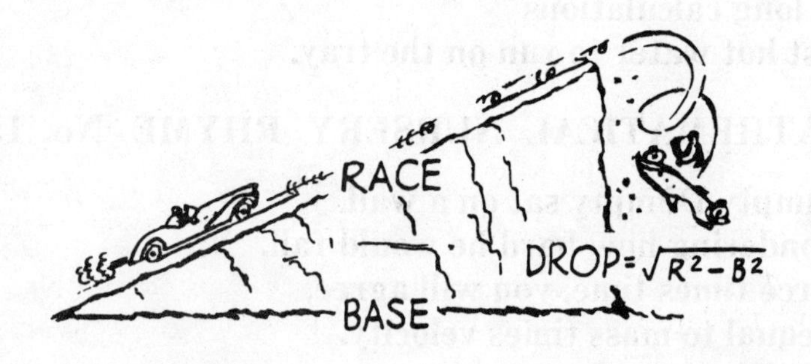

53. TRISECTORS

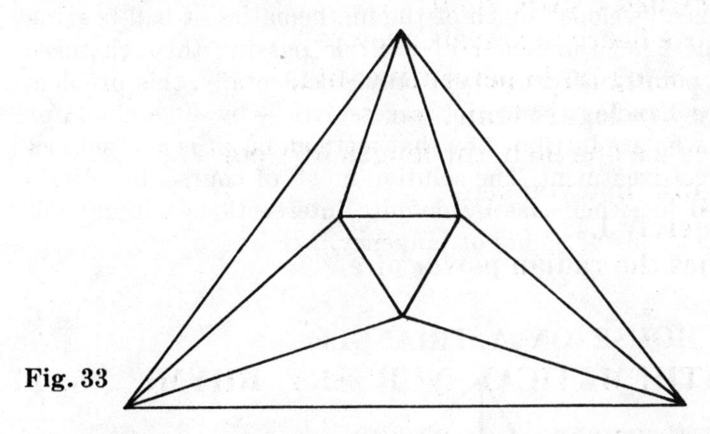

Fig. 33

Mr. H. T. Kennedy, of New York City, offers this intriguing problem in geometry: "Prove that the trisectors of the angles of any triangle intersect to form an equilateral triangle." He included this interesting commentary, "On solving the problem, some of your readers will no doubt ask the natural question that I did: Can this theorem be used as a means of accomplishing the 'impossible' trisection of the general angle? I've worked on it for four years, off and on, and my conclusion is that it cannot be so used, but I'd like to hear from your readers on the subject."

54. CENTER OF A CIRCLE

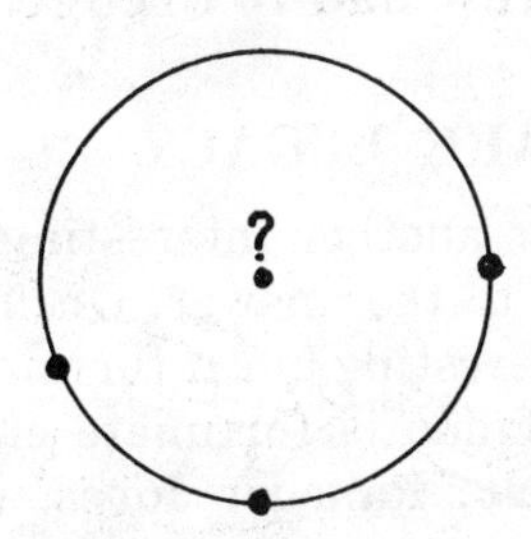

Fig. 34

This problem is similar in general character to No. 11, "The Unruly Bisection," but more difficult. It was submitted to the DIAL by A. B. Jones, then of New Departure, Bristol, who has developed much of the mathematics of ball-bearing design: Find the center of a circle passing through three given points, using compass only. Incidentally, this problem can be handled — but not necessarily — by inversion, for those who are intrigued by that method, but it is also subject to direct treatment. The solution must, of course, be accomplished in either case by definite intersections without approximations of points of tangency.

55. HOUSE ON A TRIANGLE

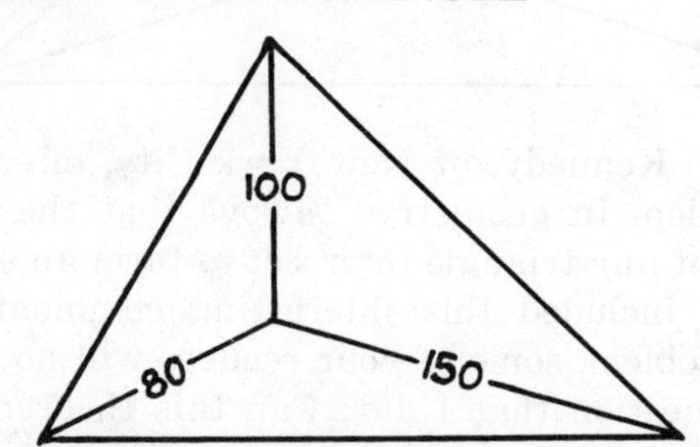

Fig. 35

Mr. E. R. Bowersox, of Los Angeles, tells of a house on a plot that was an equilateral triangle, the house being located 80, 100, and 150 yards from the respective vertices. He asks you to give the simplest way to find the length of each side of the plot, for these distances or for any distances a, b, and c.

56. FOUR-SQUARE LEGACY

This problem is another interesting example of finding the right short cut to the answer. On first review it seems laborious and uninteresting but it turns out to be a short and pleasant exercise if one is fortunate enough to hit on the symmetrical way out: Rancher Jones, who was a bit of a mathematician, left a rather unusual will. The bulk of his extensive property went to his widow but he stipulated that each of his four sons could take an exactly square plot of land, if the dimensions could fulfill these conditions: The area of John's plot expressed in square miles must equal the length of Henry's plot in miles plus two, and similarly Tom's area must equal a side of John's square plus two. Also, Henry's area added to the length of Bill's plot must equal two and likewise Bill's area plus the length of Tom's plot must equal two. What was the side of each son's square and was there any squabble because there was more than one possible settlement?

57. THE CONVEX LENS

Fig. 36

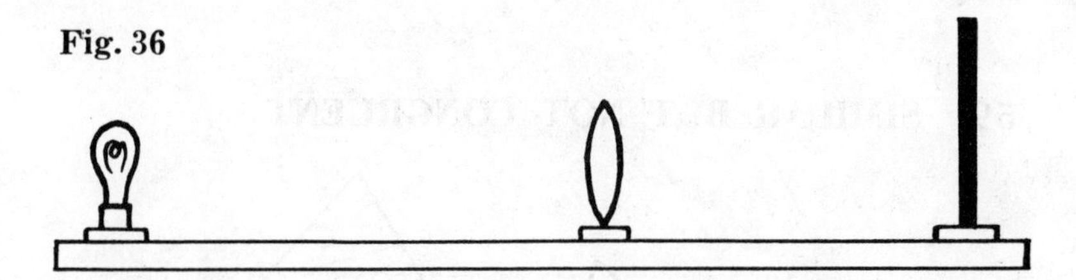

Little Euclid had advanced to optics in his physics course and was preparing to verify in the laboratory the familiar equation for a convex lens $1/f = 1/u + 1/v$, where f is the focal length, u the distance from light source to lens and v the distance from image to lens. Said the Prof., "Fasten down the lens holder here at the middle of this long shelf and then move the lamp various distances away and get the image on the screen for each setting. Measure u and v accurately on these graduated slides. Focal length of the lens is 12 cm."

"Okay," said Euclid, "but I'd like to avoid fractional measurements and have all my distances in whole numbers of centimeters."

"I don't know whether that will give enough readings," said the Prof. doubtfully.

"Oh, yes, it will!" replied Euclid brightly after a few

minutes' thought. "I'll get 15 different lamp settings to check that way."

Explain whether Euclid's figure was right and how he got it so quickly.

58. FIVES AND SEVENS

Here is another fascinating example of the magic of numbers, contributed by Mr. V. Thebault, Tennie, Sarthe, France: "If twice the square of a smaller integer is one greater than the square of a larger integer and the smaller integer ends in 5, show that the larger integer must be divisible by 7 — the same applying for 29 and 41, respectively."

59. SIMILAR BUT NOT CONGRUENT

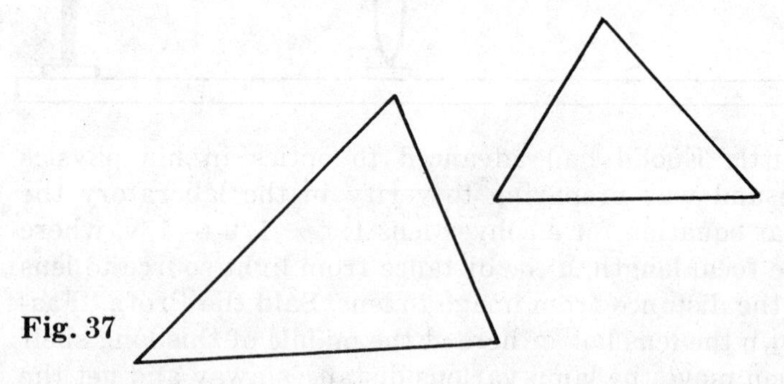

Fig. 37

Thanks also to Victor Thebault for another exploit of little Euclid, who came home one day from geometry class, where he had learned about triangles which are similar, and triangles which are congruent.

"Look, Pa," he said, "similar triangles have all their corresponding angles equal, and two triangles must be congruent if they have two sides and their included angle respectively equal. But I can draw two similar triangles right here that are not congruent even though they have two sides of one equal to two sides of the other."

How did he do it? And what is the necessary relationship for this to be true?

60. MATCHSTICK SQUARES

This interesting but not easy exercise with matchsticks was submitted by R. A. Davis, Chicago Bridge & Iron Co.,

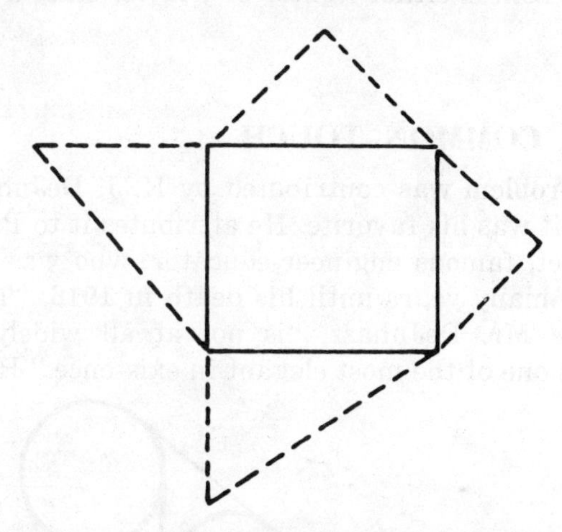

Fig. 38

Birmingham, Ala.: Using matches all of equal length, form a square. Then on each leg of the square form a right-angled triangle outside of the square; each leg of the square being one side of such a triangle and no two of the four triangles being alike. What is the fewest number of matches you will need for the job?

61. THE COUNTERFEIT COIN

This problem came to light, apparently for the first time, in 1945, when it was contributed to the DIAL in October of that year by Dwight A. Stewart, RCA, Camden. It has since appeared in various forms in many books on mathematics. It is included here chiefly because the diagram that accompanied the best solution is quite distinctive and illustrates an unusually apt approach which may be applied to other posers of this general character.

"Using a set of balances, what is the least number of weighings necessary to discover which one of twelve coins of given denomination is counterfeit, assuming that the eleven standard coins are of equal weight and that the counterfeit coin is either lighter or heavier than a standard coin?"

62. THE COMMON TOUCH

This problem was contributed by K. J. DeJuhasz, who wrote that it was his favorite. He attributes it to Prof. John Edson Sweet, famous engineer-educator, who was active at Cornell for many years until his death in 1916. "The problem," avers Mr. DeJuhasz, "is not at all widely known, though it is one of the most elegant in existence." Here it is:

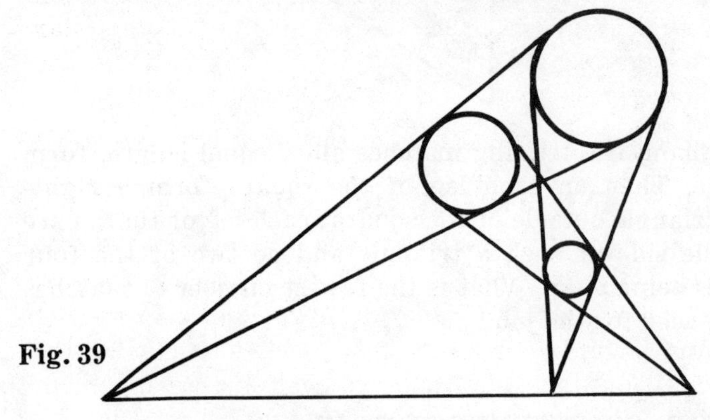

Fig. 39

Draw any three circles; show that the points of intersection of the common outside tangents to each of the three pairs of circles lie in a straight line.

63. HIGHEST APEX

Art Payne, machine designer, Bundy Tubing Company, Detroit, tells of a triangle whose base is 10 inches long and whose other two sides have lengths in the ratio of 3 to 2. On what line must the apex be located and what is its maximum possible height from the base? The solution should also include the general case where the ratio is a/b and the base unity.

64. PYRAMID OF BALLS

This tells of the problem of the advertising manager of a ball-bearing company who proposed to attract attention to his wares at a power show by building a pyramid of one-inch balls. He did not know whether to make it a triangular, square, or rectangular pyramid, but wanted, if possible, to use up all the balls he had, which amounted to 36,894. Knowing (says Mr. F. L. Nisbet, of Holley Carburetor Co., Detroit, who contributed this problem) that the construction of a triangular pyramid is such that one sphere rests on three below it, while in both the square and rectangular types, one sphere rests on four below it; just how did the ad man do the job?

65. ALICE AND THE MAD ADDER

We tell a tale of Alice and the Mad Adder that would have intrigued that famed mathematician, Lewis Carroll. It seems that Alice went to the corner butcher shop for her mother and bought four items. The clerk jotted down the four amounts but instead of adding them, in a mad moment he multiplied the four figures together and asked Alice for the money, $7.11. Alice had already made the addition mentally and since the figure checked, she paid and left. What was the cost of each of her purchases? This problem was contributed by Mr. David A. Grossman, New York.

66. THE WALK AROUND

This problem, originated by Howard D. Grossman, New York, was dubbed by one of our readers "a beautiful pursuit problem" and brings to mind some of our other problems: the ship chasing another that took refuge in a cloud bank, No. 38; our Fido-Tabby chase, No. 74; the Hunter and his

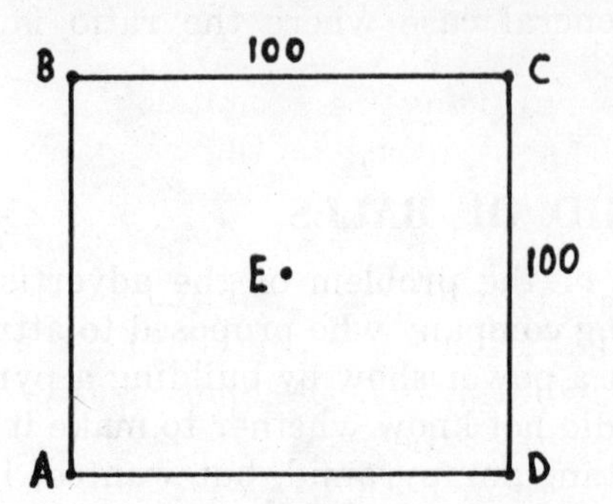

Fig. 40

Dog, No. 50, etc. In this one, little Euclid, who had been asked by the teacher to devise games at recreation period, stationed Arthur, Ben, Charles, and David at the four respective corners of the 100 ft. square schoolyard and took his position at the center of the square.

"At the signal," commanded Euclid, "you, A, start walking at this pace (suiting the action to the word) always toward B, B keep walking at the same rate toward C, C toward D, and D toward A. When you all meet here at the center, we will see how far each will have walked."

When he gave the signal, Euclid himself decided to walk at the same pace toward the midpoint of one of the edges of the yard and back, with a rather surprising result. You are asked to write the next chapter.

67. THE ABSCONDER'S LEGACY

At the risk of flopping somewhat into the flippant but by no means elementary, we present a problem submitted by Mr. C. W. Lincoln, formerly Chief Engineer of Saginaw Steering Gear Division of General Motors Corporation. Here is the gem of Diophantine devilry: "An insurance company decides to accrue a special reserve fund by transferring cash each day from its general funds into a special fund. One cent is to be transferred into the special fund on the first day after the decision is made, two cents on the following day, three cents on the day following that, and so on; the amount of cash transferred being increased by one cent each day. The fund is to be kept in cash on the company premises and no interest will accrue. The transfer is to be made at 10:00 A.M. each day. The plan is put into operation and is successfully carried out for many days, until the vice-president of the company and custodian of the special fund, enters the premises surreptitiously and absconds with the entire fund. At this time the age of the company is less than one million years. The vice-president is never caught. He makes his way with his family, to a little known Pacific island, where he makes a living for his dependents and achieves a modicum of happiness by working as a plumber's assistant. As old age approaches, the absconder decides to utilize the stolen fund, which has been kept absolutely intact, to make a gift to his grandchildren, in lieu of a legacy. There are twenty-three grandchildren and each one is inordinately fond of baboons. The grandfather decides for this reason that his gift to them will be a gift of baboons. He calls up a wholesale baboon house in a nearby metropolis and gets a quotation on baboons in quantity. The quoted price per baboon is an exact multiple of one cent. After careful calculation, the absconder finds that there is exactly the right amount of money in the stolen fund to buy for each one of the twenty-three grandchildren a number of baboons equal to the number of cents in the price of one baboon. For how many days did the insurance company transfer money into the special reserve fund?"

68. PILOTS' MEETING

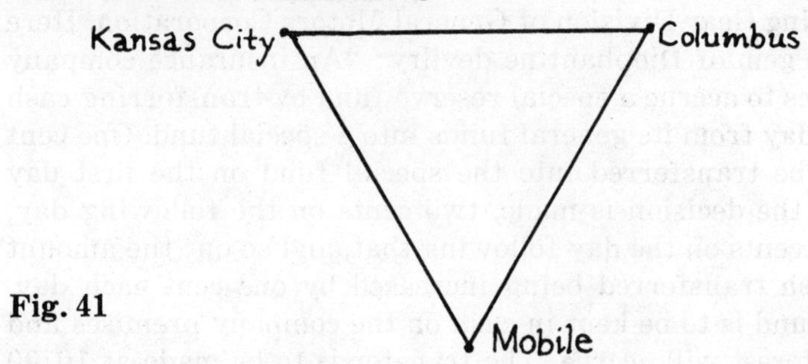

Fig. 41

This tells of three Air Force officers who found themselves respectively at points near Kansas City, Mobile, and Columbus. For the purpose of this problem, these cities can be considered 700 miles equally distant from each other. It was necessary for the three officers to get together at the earliest possible moment for an important conference. Major King's plane at Kansas City could average only 200 mph, Major Morton's craft at Columbus cruised at 400 mph and Colonel Colton at Mobile had a 600 mph jet. Colton was the mathematician and he phoned the other two and told them where to meet him. Luckily there was an airport near the desired spot. Each took off at 2 P.M. When and where did they meet. Also, if the plane speeds instead of being 200, 400, and 600 mph, are respectively 300, 400, and 500 mph; what essential difference in the situation at earliest meeting does this change produce?

69. THE CENSUS TAKER

Mr. Henry H. Rachford Jr., of Houston, Texas, informs us that little Euclid, who had grown up to be a census taker, came to the home of Pythagoras Jones. After Jones had given the requested information about himself, Euclid asked, "Who else lives here?"

Jones answers, "Three others, but none is at home."

Euclid replies, "Then perhaps you can give me their names and ages."

After giving the names in order of seniority, Jones, with a twinkle in his eye, says, "The product of their ages is 1296 and the sum of their ages is the number on this house."

Euclid ponders a moment, then asks, "Are any of their ages the same as yours?"

Jones answers, "No," whereupon Euclid leaves satisfied. What is the house number?

70 TRIANGLE OF COINS

This is a companion puzzler to our No. 30 on the Game of Nim, with some interesting variations. You may remember that in Nim, three random piles of coins are drawn from in turn by two players, each taking one or all of the coins

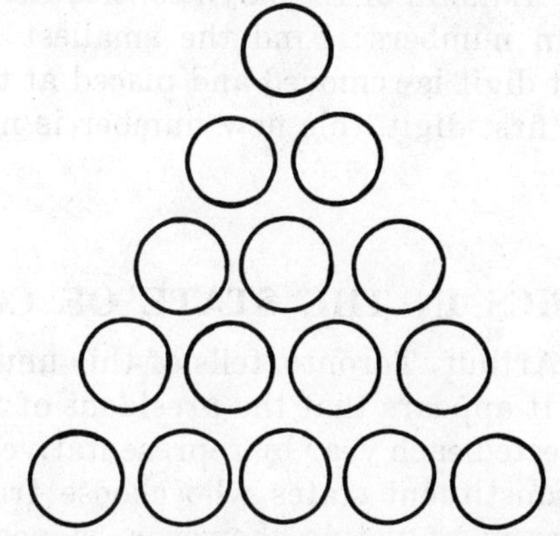

Fig. 42

in a single pile. The problem is to force the other player to take the last coin, or guarantee to take it yourself, depending on the pre-established objective. In this problem, 15 coins are arranged pyramid-fashion in five horizontal rows of 1, 2, 3, 4 and 5 each. Each of the two players in turn takes one or all of the coins in a single row, choosing any row. The game can be either to take the last coin yourself or to force your adversary to do so. If you are "in the know" and the other chap isn't, it is claimed that you can win, whoever takes first pick. How?

71. SQUARE PLUS CUBE

This fascinating example of the never-ending magic of numbers was originated by Victor Thebault, Tennie, Sarthe, France: In the simplest way you can think of, find two whole numbers which between them make use of each of the ten digits 0, 1, 2, 3, 4, 5, 6, 7, 8 and 9 just once. The two numbers are respectively the square and the cube of the same number.

72. END AT THE BEGINNING

Mr. M. S. Klamkin of Brooklyn contributes this intriguing exercise in numbers: Find the smallest number such that if the last digit is removed and placed at the beginning to become the first digit, this new number is nine times the original one.

73. POLITICS IN THE STATE OF CONFUSIA

D. S. McArthur, Toronto, tells of this unusual situation in Confusia: It appears that the president of the Confusian Republic is elected each year by representatives sent in from the various constituent states who choose from two nominees. The system of voting, however, is peculiar because although there is only one representative for each state, each representative casts as many votes for the candidate of his choice as there are representatives voting for that candidate. Due to population growth, Confusia has been divided into more states each year during the last ten years, so that the number of states has now very nearly reached the statutory maximum of 275. Strangely enough, in each of the past ten years, the majority polled by the winning candidate has always been exactly the same. How many states were there at the time of the last election?

74. FIDO-TABBY CHASE

This "Pursuit Problem" which was devised by W. W. Johnson of Cleveland as a companion to his "Euclid's Put-Put Problem" (No. 28) is perhaps of even greater interest, since in a way it epitomizes a vital difference between dumb animals and our more articulate so-called human race.. In

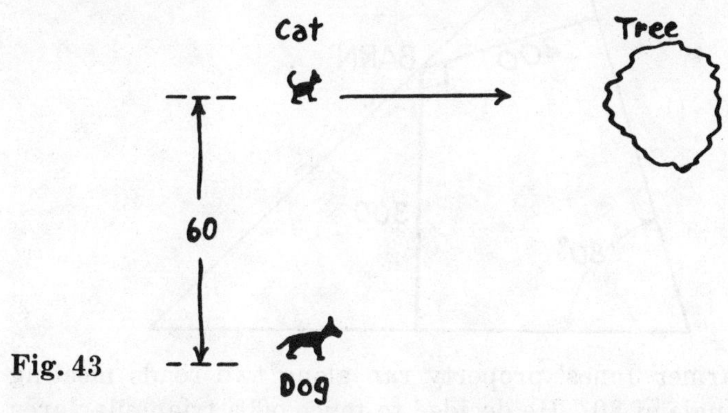

Fig. 43

this problem "a speeding dog spies a cat sixty yards due north of him and hotfooting it due east toward a tree. The dog runs one-fourth again as fast as the cat and in its eagerness to reach the prey keeps running towards it. Our question is — how far does the cat go before being caught?" Now, we mention in passing that if the dog had been Euclid metempsychosed and knew that a straight line was the shortest distance between two points and if he had instantly remembered with Pythagoras that in the familiar right triangle with sides of 60, 80, and 100 the hypotenuse happens to be 25% longer than the longer leg; he would have promptly set his course at N. $\sin^{-1} 4/5$ E., and would have caught the prey handily after it had run only 80 yards. But sad to say, Fido, not possessed of human knowledge and foresight, must waste time and effort by pointing constantly toward the cat and failing to intercept Tabby at all before she reaches the haven of the tree. At any rate, that is for you to discover.

75. THE TRIANGULAR FENCE

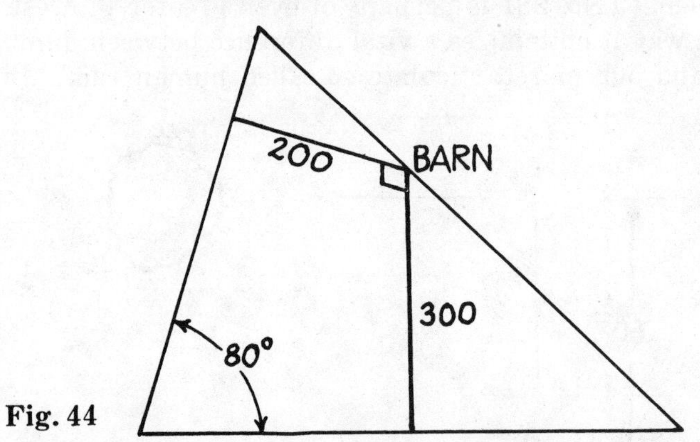

Fig. 44

Farmer Jones' property ran along two roads meeting at an angle of 80°. He decided to fence off a triangular area bounded by the two roads, with the third side passing just back of his barn, which was 200 feet from one road and 300 feet from the other. In talking things over with his young son, who was none other than our little Euclid, the hero of many previous problems, he remarked that he might as well use up the entire 2000 feet of fencing that he had in the shed. "That's easy," said Euclid, "I'll show you how to locate that third side of the triangle in two shakes," which he did. How?

MATHEMATICAL NURSERY RHYME No. 18

"Column, column in the wall,
How much load before you fall?"
"My buckling day is quite afar
If you don't exceed my l/r."
"Beam, beam in the floor,
When will you carry load no more?"
"My useful days will cease to be
When stressed beyond my M/Z."

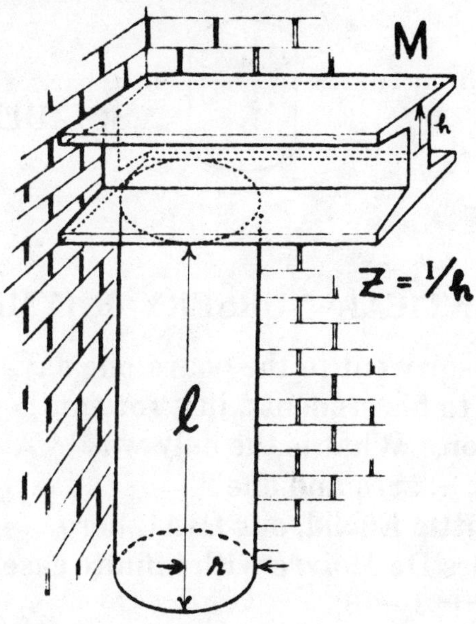

MATHEMATICAL NURSERY RHYME No. 19

See Saw, Marjorie Daw,
She rocked — and learned the lever law.
She saw that he weighed more than she
For she sat higher up than he,
Which made her cry, excitedly,
"To see saw, it is plain to see,
There must be an equality
'Twixt You times X and Y times Me."

MATHEMATICAL NURSERY RHYME No. 20

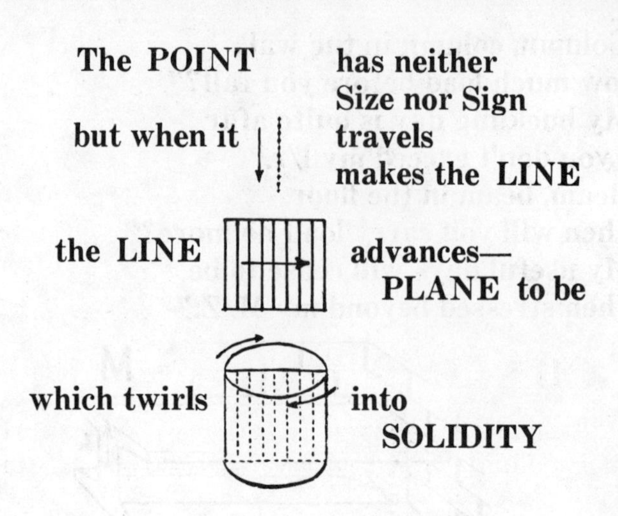

The POINT has neither
Size nor Sign
but when it travels
makes the LINE
the LINE advances—
PLANE to be
which twirls into
SOLIDITY

MATHEMATICAL NURSERY RHYME No. 21

The professor went to the board one day
And posed to his students, just for fun,
The question, "What is the only way
To link e, i, π, zero and one?"
Up spake little Euclid, our DIAL hero,
Who handles De Moivre with infinite ease,
"Sure, $e^{i\pi} + 1 = 0$,
Give us a hard one. Next question, please!"

25 CHALLENGING QUICKIES

THESE ARE SELECTED for their special interest from the "Companion Puzzlers" that accompany the longer problems in most issues of the DIAL. Readers are invited to keep track of the time taken to reach the solution to these quickies, which should not normally exceed five minutes, since most of them are chiefly a test of mental agility in finding the best short-cut to the answer.

76. BALLS IN THE BOX

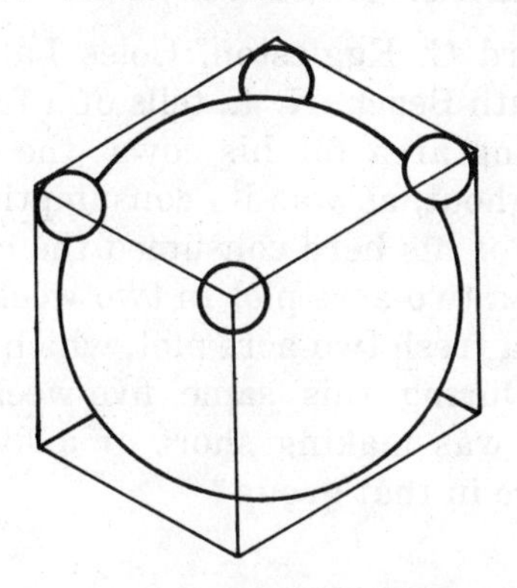

Fig. 45

A manufacturer of steel balls shipped his 2″ balls individually in a lined cubical box with inside dimensions of substantially 2″ to assure a snug fit. It occurred to him that he might use the space at the top four corners to carry four additional balls that would fit snugly against the big ball, two sides of the box and the box top. What would be the approximate diameter of the four small balls?

77. THE FOUR-CUSHION SHOT

Murray S. Klamkin, Brooklyn, tells of a billiard player who hit a ball with no "English" in such a way that it returned to its starting point after hitting four different cushions. How far did it travel?

78. SUM OF THE CUBES

Dr. Ernest Rabinowicz, M.I.T., invites you to find quickly any two different integers, the sum of whose cubes is the fourth power of an integer. And, after finding one pair of numbers that fill the bill, tell how you could find all such cases.

79. THE GRAZING COWS

Mr. Richard C. Eggleston, Coles Laboratories, Signal Corps, Monmouth Beach, N. J., tells of a farmer who had an extensive grazing area for his cows, the grass yield being uniform throughout, as was its consumption by each of the animals. Eight of his herd consumed the original grass and new growth on a two-acre plot in two weeks, after which he moved them to a fresh two-acre plot, which lasted them three weeks more. During this same five-week period another group of cows was making short of a five-acre plot. How many cows were in that group?

80. THE WOBBLY WHEEL

Mr. David A. Lee, AiResearch Manufacturing Company, Los Angeles, Cal., tells of a mechanic who had a wheel

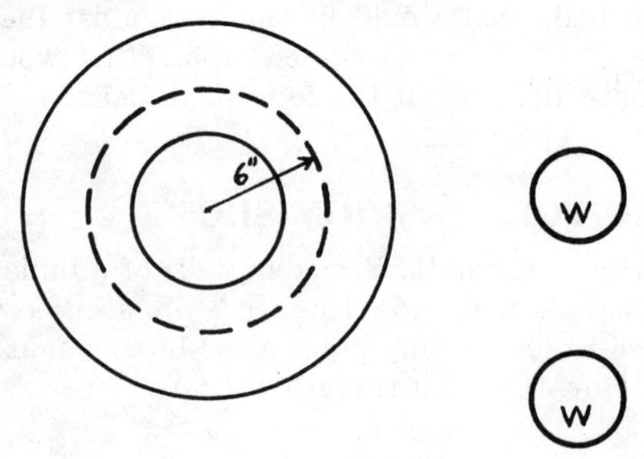

Fig. 46

that was pretty badly out of balance. All he had on hand to reduce the unbalance was two equal weights, each weighing an integral number of pounds. Each weight could be fastened at any desired point six inches from the center of the wheel, whose shaft would be supported on knife edges while making the static correction. At what two points should the weights be fastened, one after the other, to do the job? And if the unbalance were six inch-pounds and the weights two pounds each; how far apart would they be placed?

81. OLD MAN AND HIS SON

W. W. Johnson, of Cleveland, writes, "An old man and his son make a 64-mile trip with their saddle horse which travels 8 miles per hour but can carry only one person at a time. The man can walk 3 miles in one hour and the son 4 miles. They alternately ride and walk. Each one ties the horse after riding a certain distance, and then walks ahead, leaving the horse tied for the other's arrival. At the halfway point they come together and take one half-hour for lunch and to feed the horse. At what time do they arrive at their destination if they start at 6 A.M.?"

82. SHEET OF PAPER

A. W. Everest, General Electric Co., Pittsfield, Mass., invites you to cut a sheet of typewriter paper into three pieces which can be rearranged into a square; prove your solution.

83. EARLIEST TIE FOR THE PENNANT

In May, 1955, when the Dodgers had won so many of their early games as to make it seem a runaway race for the National League pennant, Mr. D. R. Chicoine, Project Engineer, A. E. Staley Mfg. Co., Decatur, was inspired to make this provocative query: "What is the earliest point in a regulation major-league baseball schedule that a team can be assured at least a tie for first place in the final standings?"

84. ALL FOUR AILMENTS

This one, contributed by Norman Anning, Ann Arbor, is a rather unique test of alertness of thought: "If 70% of the population have stomach trouble, 75% have weak eyes, 80% have a liver ailment, and 85% have a touch of catarrh; what per cent at least are plagued with all four ailments?"

85. DELETED CHECKERBOARD

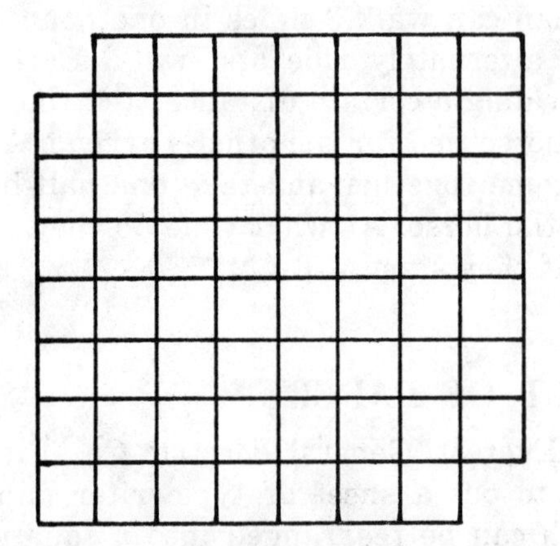

Fig. 47

Robert S. Raven, Westinghouse, Cleveland, offers this unique poser: "Two diagonally opposite corner squares are cut out from a checkerboard. The checkerboard now has an area of 62 squares. You are supplied with 31 identical dominoes with dimensions two squares by one square. The total area of the dominoes is thus also 62 squares. It is now proposed to cover the deleted board completely with non-overlapping dominoes. Prove this is possible or impossible in one or two sentences."

MATHEMATICAL NURSERY RHYME No. 22

Tom, Tom, the piper's son,
Could pipe as pert as any one
In longitudinal waves of strength
With frequency inverse to length.
And every blast, when he'd begin it,
Blared out at thirteen miles a minute.

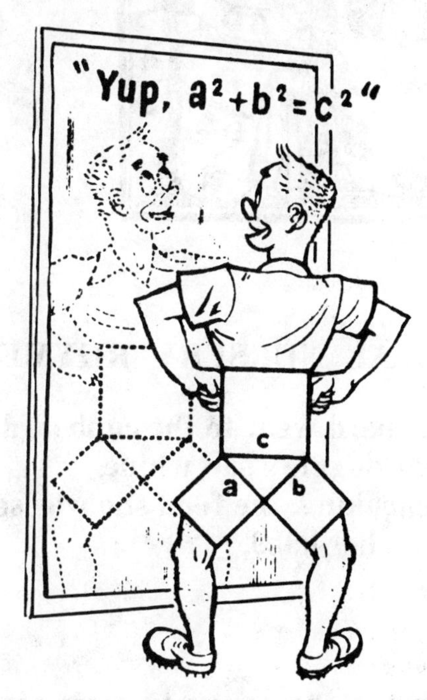

MATHEMATICAL NURSERY RHYME No. 23

If a pair of shorts you need
Pythagoras is the man to heed.
Layout the squares per sketch above;
If you like Euclid, this you'll love.
Then simply cut the pants to fit. —
You'll be the latest fashion hit.

MATHEMATICAL NURSERY RHYME No. 24

"O shade of Bernoulli , tell me truly,
How did your numbers grow?
"With B's and C's and a sprinkle of e's
And factorials all in a row."

MATHEMATICAL NURSERY RHYME No. 25

Old Mother Hubbard went to the cupboard
To stop her poor doggie's sad whine.
Algebraically speaking, the food she was seeking
Made the look on her k9b9.

MATHEMATICAL NURSERY RHYME No. 26

Fiddle de dum, fiddle de dee,
A young man sat beneath a tree
And pondered over gravity:
"Cm sub one times m sub two
Divided by d squared, 'tis true,
Denotes the force that follows through."
Then m sub one hit youth on dome
And gave conviction to this pome.

86. THE PLATE OF PIE

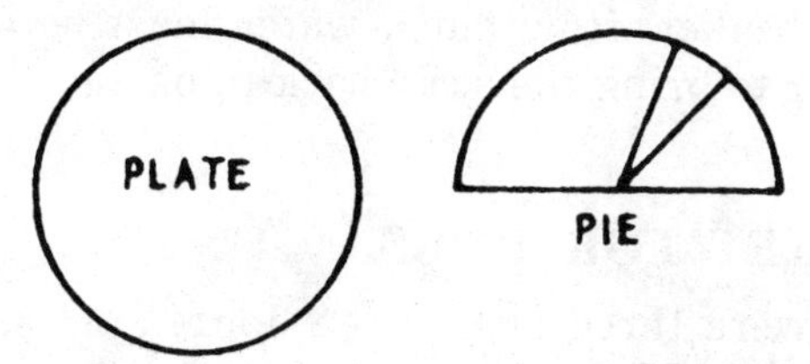

Fig. 48

Viktors Linis, Saskatchewan, wants to know the diameter of the smallest circular plate on which a semi-circular pie can be placed if the pie is cut in sectors of the same radius as the pie?

87. ROD IN THE BEAKER

M. F. Thomas, Socony Vacuum Oil Co., N. Y., contrib-

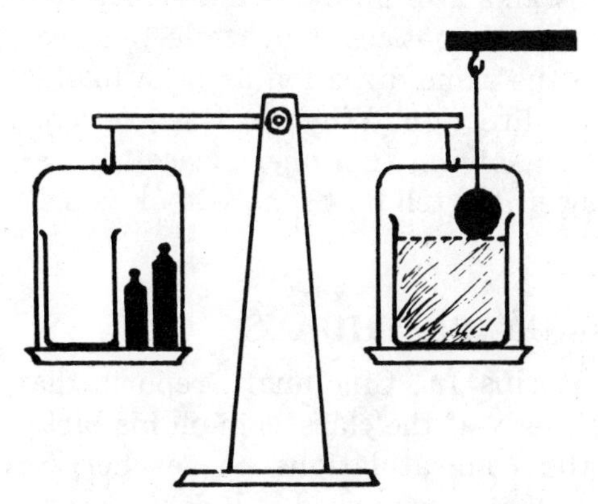

Fig. 49

utes this one: A cylindrical iron bar, one square centimeter in cross-section, is suspended vertically over a beaker, two square centimeters in cross-section and partly filled with water, so that the bar just touches the surface of the water. The beaker stands on one pan of a balance scale; the other pan carries an empty beaker and sufficient gram weights to balance the scale. One cubic centimeter of water is added to the first beaker. How much water must be added to the other beaker to bring the scale back to balance?

88. MARKET FOR HOGS

There were three men called John, James, and Henry and their wives, Mary, Sue, and Ann, but not respectively. They went to market to buy hogs. Each person bought as many hogs as that person spent dollars per hog. John bought 23 hogs more than Sue; and James bought 11 more than Mary. Each man spent $63 more than his wife. What was the name of each man's wife?

89. PASSING STREETCARS

A man walks at 3 mph down a street along which runs a streetcar line. He notices that, while 40 streetcars pass him traveling in the same direction as he walks, 60 pass him in the opposite direction. What is the average speed of the streetcars? In solving this one, please give adequate proof of your answer and tell how long it took you.

90. COMMON BIRTHDAYS

A. H. Phelps Jr., Cincinnati, reports that little Euclid arrived smilingly at the classroom on his birthday and after receiving the congratulations of teacher Smith and the other 29 students, remarked, "I don't know if any other boy in the class has his birthday today but I'll bet ten cents to four bits that there are at least two boys in the room who celebrate their birthdays on the same day, whenever that may be." Mr. Smith, after mentally setting up the fraction 30/365 thought that Euclid was giving generous odds and that he would be sadly out of pocket if he went through the school, which had a couple dozen classrooms of that size, making such a bet. Just how foolish was our little Euclid?

91. RESISTANCE 'CROSS THE CUBE

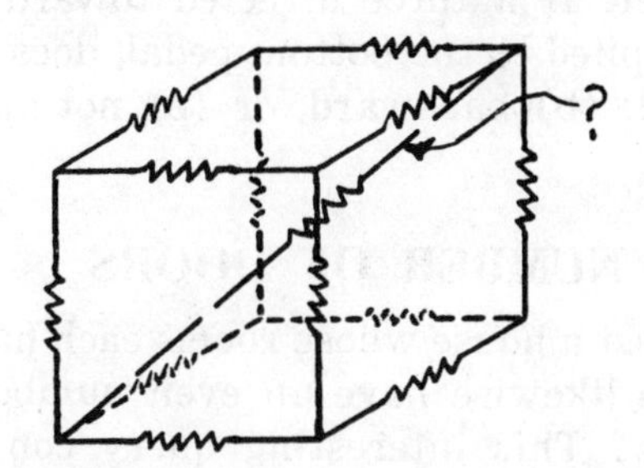

Fig. 50

A 1″ cube has its outline constructed out of wire whose resistance is one ohm per inch. What is the resistance between opposite corners of the cube?

92. MEDIANS AND OLDHAMS

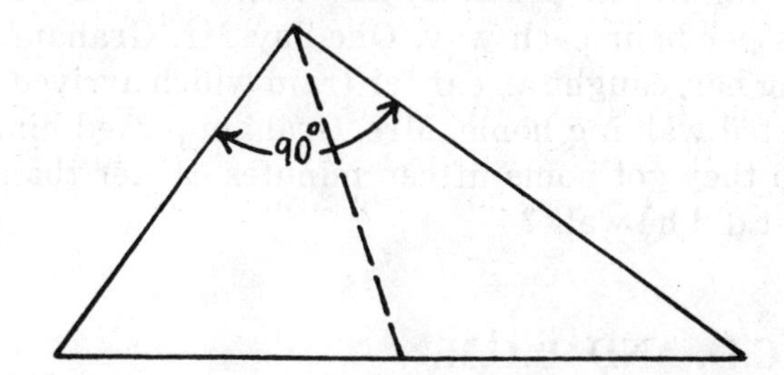

Fig. 51

What is the quickest way to prove that the median to the hypotenuse of a right triangle is equal to half the hypotenuse? And, from a practical angle, if you happen to be familiar with shaft couplings, what has this got to do with an Oldham coupling?

93. THE BACKWARD BICYCLE

Try your hand at this "experiment in mechanics" devised by Prof. Ernest Rabinowicz, M.I.T.: The pedals of an ordinary bicycle are so positioned that one is below the other and the saddle is held lightly to keep the bicycle from toppling over. If a force directed toward the rear of the bicycle is applied to the bottom pedal, does the bicycle move (a) forward, (b) backward, or (c) not at all? Why?

94. EVEN NUMBER OF DOORS

Why must a house whose rooms each have an even number of doors likewise have an even number of outside entrance doors? This interesting query comes from Roland Silver, Electrical Engineer, Electronics Corp. of America, Cambridge, Mass.

95. THE AMBULATORY COMMUTER

Mrs. Suburban Graham met her husband at the station with the family car promptly at 5 P.M. every day, averaging 30 miles per hour each way. One day Mr. Graham, without notifying her, caught an earlier train which arrived at 4 P.M. and started walking home. Mrs. Graham picked him up part way and they got home fifteen minutes earlier than usually. How fast did he walk?

96. G.C.D. AND L.C.M.

Harvey Berry of Lexington, Ky. asks you to quickly prove that the G.C.D. of two numbers is equal to the G.C.D. of their sum and their L.C.M.

97. THE TRACKWALKERS' QUANDARY

Mr. M. H. Kapps, in submitting this Quickie, wrote: "I have one that is not difficult but requires a little straight thinking, which your readers would undoubtedly appreciate, so I am forwarding it to you: Two men are walking along a railroad track. A train passes the first man in ten seconds. Twenty minutes later it reaches the second man. It passes the second man in nine seconds. How long after the train leaves the second man will the two men meet (all speeds constant)? Direction and speeds of travel, train length, and distances are not given nor are they required."

98. CONSECUTIVE INTEGERS

Mark Holzman, Res. Engr., Western Geophysical Company, invites you to find three consecutive integers such that; when all possible simple fractions are formed therefrom, the sum of the six different fractions will be an integer.

99. COWS, HORSES, AND CHICKENS

Roger Salmon, Fullerton, Cal., tells of gentleman farmer Jones who was taking count of his horses, cows, and chickens. Being somewhat of a mathematician, he noticed that he had a different prime number of each. Moreover, he observed, "If I multiplied the number of cows I have by my sum total of cows and horses, it would give me a figure just 120 greater than my number of chickens." How many of each must he have had?

100. BINGO CARDS

Alice Anderson, Knoxville, asks you to tell how many different Bingo cards can be made in which no card can have a row, column, or diagonal with numbers identical to those in a row, column, or diagonal of another card; there being five numbers in each column selected respectively from 1 to 15, 16 to 30, 31 to 45, 40 to 60, and 61 to 75, except that the space common to the third column and the third row is blank.

SOLUTIONS TO PROBLEMS

WHEN THESE PROBLEMS originally appeared, prizes were given for the best (and, usually, shortest) solution, and if there were several differing modes of solution of about equal merit, additional consolation prizes were awarded.

Unlike the procedure that must be followed in textbooks, we have frequently included in the pages that follow a number of different solutions for an individual problem and have usually given the answers in substantially the words as submitted. In this way the reader is enabled to compare his own solution not only with the best answer, but with others that may parallel his own.

1. THE PLUG GAUGES

This innocent-looking problem, like many in this book, can be solved in several different ways; one of which requires more originality or "mathematical ingenuity" and therefore is shortest and best. The approach here can be simple geometry, trigonometry, analytical geometry, or inversion. Of these, the simplest is geometry, as applied in the upper diagram by E. A. Terrell, Jr., of Terrell Machine Co.,

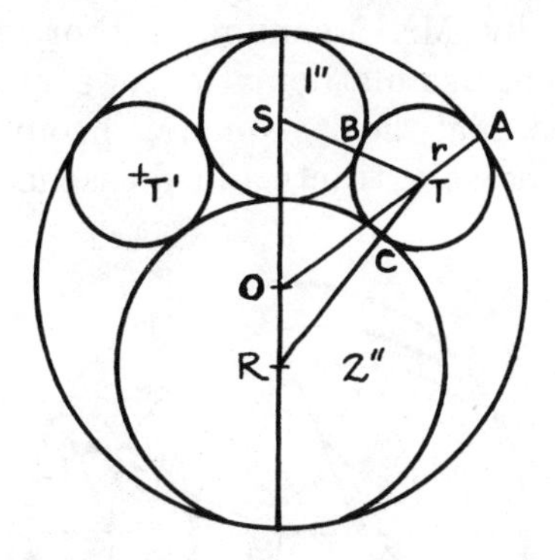

Fig. 52

Charlotte, N. C. The problem, of course, is to find the size of circle that is at the same time tangent to the outer circle and to each of the two inner circles. Assuming that such a circle has been found with radius AT equal to r, then OT is $3/2 - r$, RT is $1 + r$, and ST is $1/2 + r$. Note that triangles STO and OTR have the same altitude but, since $SO = 1$ and $OR = 1/2$, the area of STO is twice that of TOR. Using Hero's familiar formula $A = \sqrt{s(s-a)(s-b)(s-c)}$ for the area of a triangle in terms of its three sides and s, the semi-perimeter, we find that for the larger triangle s is $1/2\ \{1 + (1/2 + r) + (3/2 - r)\}$ or 3/2 and for the smaller triangle it is $1/2\ \{1/2 + (3/2 - r) + (1 - r)\}$ or the same figure, 3/2. The area relation between the two triangles then gives the equation $(3/2)\ (1/2)\ (1 - r)\ (r) = 4\ (3/2)\ (1)\ (2)\ (1/2 - r)$, from which $1 - r = 4 - 8r$ and $r = 3/7$ so that the diameter of each of the two required gauges is 6/7".

A similar solution making routine use of trigonometry applies the law of cosines, the cosine of the angle SRT being expressed successively in terms of the length of the sides of triangles STR and OTR and the resulting equation solved for r. Actually, the arithmetic becomes more involved in this method, since it brings in the square of binomials. Thus the trigonometric approach is not as simple as the one above, which makes adroit use of the common altitudes of the two triangles.

The solution by Mr. Varnum, author of the problem, though it does not use mathematics as elementary as either of these, applies analytical geometry in an interesting and ingenious way and so extracts added essence from the prob-

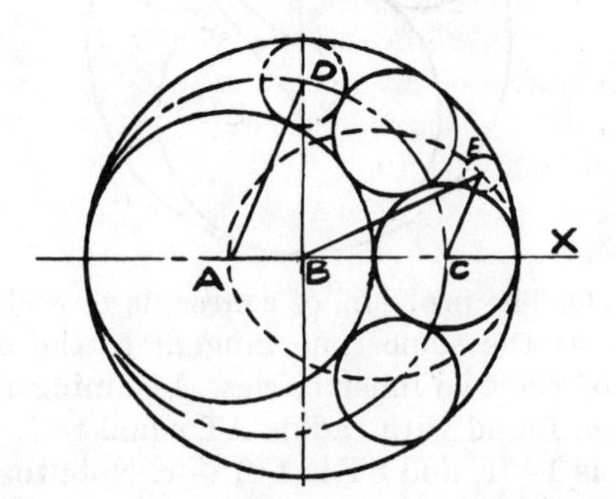

Fig. 53

lem. His mode of attack is to find the locus of the center of all circles tangent to the outer circle and the large inner circle and then to find the locus of all circles tangent to the outer circle and the small inner circle and "merge" the two loci. The center of a typical circle with radius m of the first variety is D (see Fig. 53), which is distant $(1 + m)$ from A and $(1\text{-}1/2 - m)$ from B; the sum of the two distances thus always being 2-1/2. Similarly, for circles of the second variety with radius n, one having its center at E, the distances from E to B and C are respectively $(1\text{-}1/2 - n)$ and $(1/2 + n)$; the sum always being 2. The two loci are therefore ellipses (by definition of an ellipse), as shown in Fig. 53, with equations respectively

$$\frac{(x + 1/4)^2}{25/16} + \frac{y^2}{3/2} = 1 \text{ and}$$

$$(x - 1/2)^2 + \frac{y^2}{3/4} = 1$$

Solving these equations gives their intersection at point (9/14, 12/14), the required diameter being 6/7″, as before.

2. LARGEST PRODUCT

This problem can likewise be solved in any one of a dozen possible ways — but in only one good way.

The unimaginative procedure is to laboriously set down one number as 1000a + 100b + 10c + d, and the other as 10,000e + 1000f + 100g + 10h + i, multiply through and then analyze each individual cross product, starting with the highest. Another possibility is to use the analogy of a rectangle, with each side successively increased to form the partial products (see Fig. 54 not to scale), realizing that

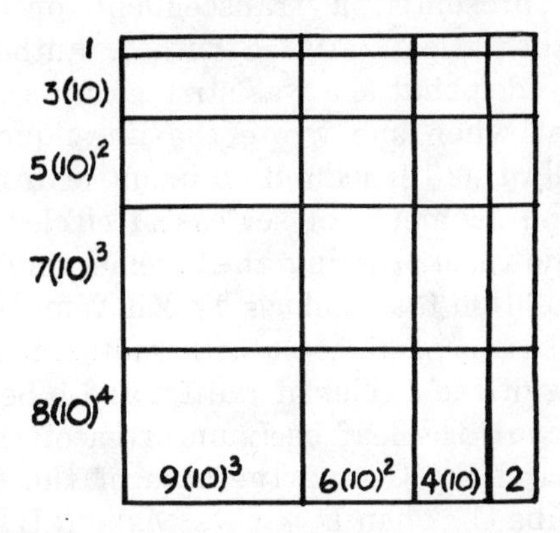

Fig. 54

area is greatest when the added rectangles at each step give a total rectangle closest to a square. A third method is to consider the logarithms of the two multipliers, which is similar in reasoning but clumsier in execution than the optimum approach.

Clearly, the simplest solution to which all the others essentially reduce is to apply the familiar rule that if the sum of two numbers is constant, their product increases as their difference decreases. It will be recalled that we readily proved this in our high-school algebra days by letting $x + y = k$, $x - y = d$, squaring both sides and subtracting, from which $4xy = k^2 - d^2$. Since it is evident that the larger digits must be as far to the left as possible, we readily set down the following pairs in succession, applying the rule above, which means that the smaller of the two added digits at each step attaches to the larger number, so as to keep the difference between the two at a minimum. Note that the final "1" must likewise annex to the smaller number:

9	96	964	9642	9642
8	87	875	8753	87531.

3. MRS. MINIVER'S PROBLEM

Mrs. Miniver's enigma of the life shared in the overlapping circles presented a transcendent problem; both social and mathematical. Miss Struther, author of *Mrs. Miniver,* had no idea that she was stirring up a mathematical hornets' nest when she wrote the lines quoted in the problem. Actually this turned out to be quite an interesting application of the geometry of sectors of circles and transcendental equations, expressing their relationships, as will be seen in the solution that follows by Mr. Wm. W. Johnson, of Cleveland:

"Let the areas of the circles of radii a and b be A_1 and A_2 $(a \leq b)$, the area of the leaf be L, the area of the crescent of the smaller circle be C_1 and the area of the crescent of the larger circle be C_2. Then $L + C_1 = A_1$ and $L + C_2 = A_2$ whence $2L + C_1 + C_2 = A_1 + A_2$. For the perfection Mrs. Miniver envisions $L = C_1 + C_2$ whence $3L = A_1 + A_2$ and

$L = (A_1 + A_2)/3$. As seen in Fig. 55B, the limit of L is A, whence $A_1 \leq (A_1 + A_2)/3$ and therefore $A_2 \leq 2A_1$. Thus it

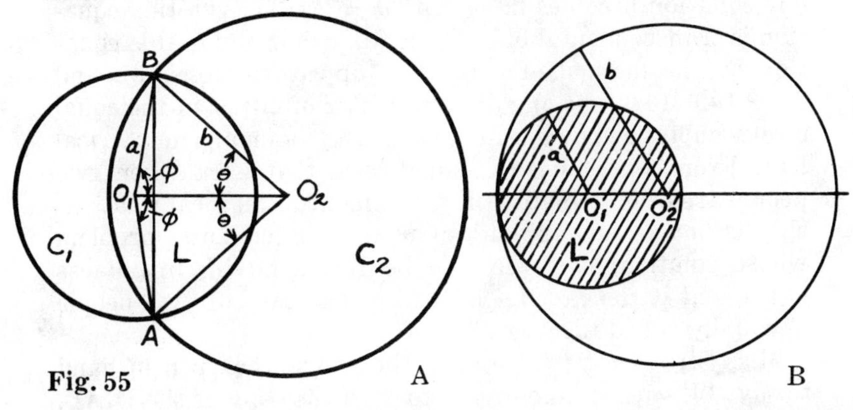

Fig. 55 A B

is seen that if the capacity of the more capable of the two individuals concerned is more than twice that of the less capable, no ideal relationship can be realized. In case it is exactly twice, the less capable will be completely immersed in the other and no private resources will be left to that member of the pair. In the general case let the ratio of the larger to the smaller radius be $r = b/a$; then $r \leq \sqrt{2}$ for the ideal condition to exist. If one arc side of the leaf subtends an angle 2θ at the center 0_2 (Fig. 55A) of the larger circle and the other arc side of the leaf subtends an angle 2Φ at the center 0_1 of the other circle, then it is seen from the figure that L = sector 0_2AB − triangle 0_2AB + sector 0_1AB − triangle 0_1AB, so that for ideal conditions when $3L = A_1 + A_2$, $L = b^2\theta + a^2\Phi - a \sin\theta\,(a \cos\Phi + b \cos\theta) = \pi\,(a^2 + b^2)/3$. But $b \sin\theta = a \sin\Phi$, whence $a \cos\Phi = \sqrt{a^2 - b^2\sin^2\theta}$, so that $b^2\theta + a^2\,\mathit{arcsin}\,(b \sin\theta) - a \sin\theta\,(b \cos\theta + \sqrt{a^2 - b^2\sin^2\theta}) = \pi\,(a^2 + b^2)/3$; or since $b/a = r$, the equation in terms of the unknown angle θ is $r^2\theta + \arcsin(r \sin\theta) - \sin\theta\,(r \cos\theta + \sqrt{1 - r^2\sin^2\theta}) = \pi\,(1 + r^2)/3$. This is the equation that must be solved for θ when the ratio r is given. In a given case this transcendental equation could only be solved by trial involving much time and patience. Thus instead of a neat mathematical formula for the solution, mathematics remains true to the social problem by presenting a situation of transcendent difficulty for an ideal solution. In the special case of a perfect mating, $r = 1$ and our equation becomes $2\theta - \sin 2\theta = 2\pi/3$. Even this equation is transcendental but not too difficult. Solving this equation by the method of successive approximations, we find $2\theta = 149° 16.3'$, the angle at the center of either of the equal circles subtended by one arc side of the common symmetrical leaf. From the above we can deduce that even when two people are exactly mated the attainment of an ideal relationship is not a mundane affair but a transcendental problem whose solution, however, can be realized with much less effort and patience than is true in the case of two people unequally yoked together."

Miss Struther, after reading the above, took pen in hand to say: "It was so kind of you to send the copy of the DIAL. I enjoyed looking at it, and I was interested to see the solution of Mrs. Miniver's problem. I'm afraid the mathematics are a bit highbrow for me, but I'll take your word for them."

4. THE BRIDAL ARCH

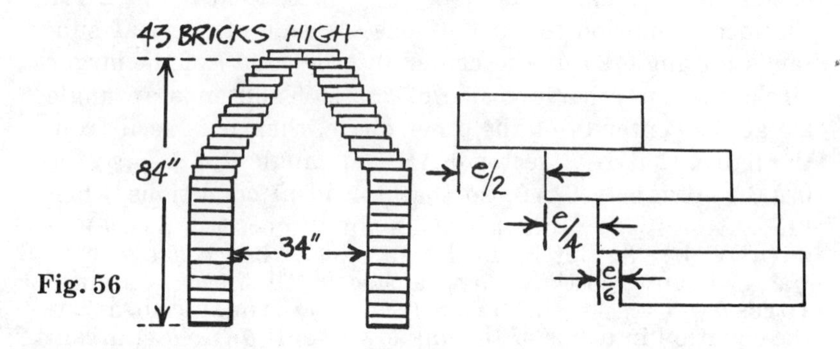

Fig. 56

This problem brought forth some architectural designs that were marvels of graceful instability. We reproduce a few of these as illustrations of how a simple and straight problem is often approached by circuitous routes. Actually, there is only one solution that provides a simple, sound, archlike arch. Clearly, the bricks should be laid on their 4 x

8-inch faces with the 8-inch dimension parallel to the plane of the arch. Since the arch would be symmetrical, consider each half made up of 43 bricks (this checks with the requirement of 7-ft. height) laid one above the other. The top brick could overhang the second by half its length without toppling, and the two top bricks by consideration of center of gravity, could overhang the third by a quarter-length; the three top overhang the fourth by a sixth-length, etc. till the forty-second brick topped the bottom brick by 8/84 inch. This gives an allowable distance between the two bottom bricks of $4(1 + 1/2 + 1/3 + 1/4 \ldots 1/42)$ which, by adding up the first 42 figures in a handy table of reciprocals, is in excess of the prescribed minimum of 34″. Actually, the first 39 reciprocals times four gives slightly more than 34″, so the simplest procedure is to lay the lowest four bricks in each half vertically above each other and 34 inches apart and then overhang the next brick in each half by 4/38″, the next by 4/37″ and so on until the two top bricks meet, with a slight fraction of safety to spare. Exactly this procedure was followed by John F. Moore, Raytheon Mfg. Co., Newton, Mass., whose sketch is reproduced at the left in Fig. 56. Harry Ellerman of Westinghouse, Ordnance Engineering, Sharon, Pa., and Fred W. Hannula, The Foxboro Company, Foxboro,

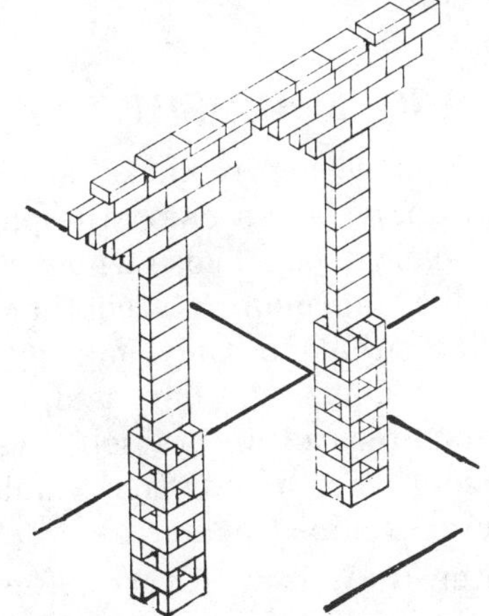

Fig. 57

Mass., also remind us that the familiar harmonic series within the parentheses is divergent (as is readily proved by taking successive groups of 1, 2, 4, 8, 16 terms etc., each of the group obviously totaling more than 1/2), whereby the width of arch with an infinite number of brick so laid could actually be infinite; a fact not readily apparent to the lay layer.

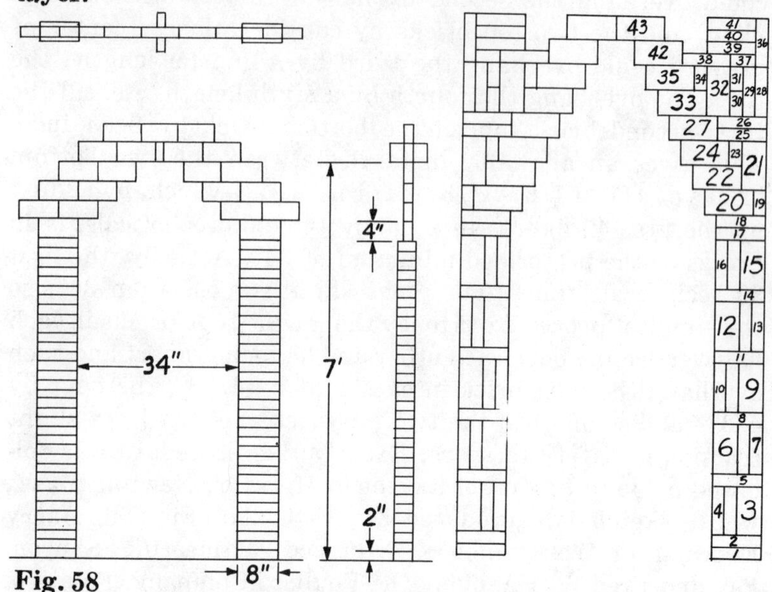

Fig. 58

5. THE CREEPING ROOF SHINGLE

This problem is a rather remarkable example of how pure mathematical analysis can chart the path to an unexpected and unusual conclusion. Those who were able to solve it commented on this and many stated that it was one of the most fascinating problems they had ever encountered. It was little short of amazing, they said, to learn that a cyclic change in temperature can in itself create the equivalent of continuous motion of a body that would otherwise be held by friction to an inclined plane.

Wrote Mr. Walter R. Wilson, Electric Research Section, General Electric, Pittsfield, Mass., whose solution was one of the most complete: "If the shingle and roof be perfectly flat, if the temperature does not drop and then rise again at any time during the day, and if the laws of friction hold for extremely low velocities; the shingle in this fascinating problem wll creep down the roof 2.2 mils during the day.

"The first point to establish in the solution is that the combination of gravitational and expansion forces cannot cause the shingle to slide as a body. If the shingle were to start sliding, the frictional force holding it back would be Wf cos $\theta = 2$ (.5) (.940) $=$ .94 lbs.

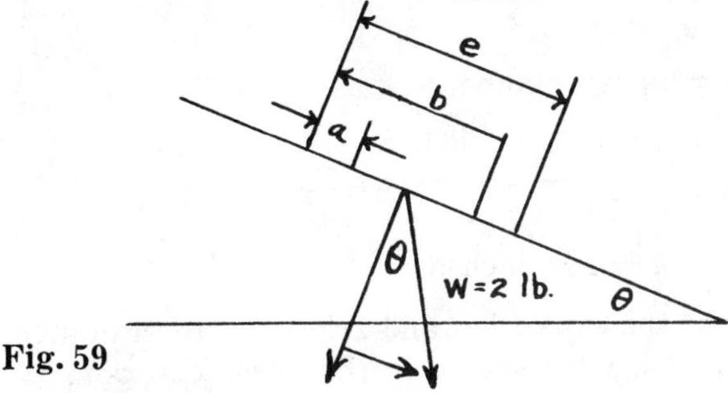

Fig. 59

"However, the component of gravitational force down the slope is only W sin $\theta = 2$ (.342) $=$ .648 lbs. The forces due to the velocity of the expanding or contracting shingle play no part because the frictional force is independent of velocity. Therefore, the sum of the forces is insufficient to maintain sliding even for an instant, and the shingle cannot slide any distance as a body.

"This means that as the shingle expands or contracts one point must remain stationary. Let a be the distance (Fig. 59) from the top end of the shingle to this point. Because the magnitude of the frictional force is independent of speed and because the direction is opposite to motion, during expansion the frictional force exerted by the roof on the portion of the shingle above the stationary point is $-(\text{Wf} \cos \theta)\,\dfrac{a}{e}$ (positive direction down slope). The frictional force on the remainder of the shingle is $+(\text{Wf} \cos \theta)\,\dfrac{e-a}{e}$. Writing Newton's law, force equals mass times acceleration of the center of gravity, $(\mathrm{WF}\cos\theta)\dfrac{a}{e} - (\mathrm{WF}\cos\theta)\dfrac{e-a}{e} +$ $\mathrm{W}\sin\theta =$ (mass) (acceleration).

"Obviously the total motion of the center of gravity is only a few miles in several hours, so the right member of the above equation can be set equal to zero (it is negligible in comparison with the one known term W sin θ). Dividing by Wf cos θ, $\dfrac{e-2a}{e} = \dfrac{\tan\theta}{f}$.

"Substituting known numbers,

$$\frac{10-2a}{10} = \frac{.364}{.5} = .728$$

$$a = 1.36 \text{ inches}$$

"The action of the expansion and frictional forces during this part of the cycle are shown in the lower diagram.

"On contraction, the signs of the two friction terms are reversed. The distance from the top to the stationary point is b. Repeating the above procedure,

$$\frac{2b-e}{e} = \frac{\tan\theta}{f}$$

$$b = 8.64 \text{ inches}$$

The distance between b and a is b − a = 7.28 inches.

"On expansion, point a remains fixed to the roof and point b creeps forward. On contraction, point b remains fixed so the total displacement of the shingle when it returns to its original size is the peak expansion of the distance between b and a; i.e., 7.28 (6) 10^{-6} (50) or 2.2 mils."

An interesting variation in approach, which considered the motion of the center of the shingle and thereby reduced the algebra somewhat, was contributed by Mr. E. F. Yendall. It is worth quoting his answer in full: "The shingle will creep down the roof .00218″ per day.

"The force pressing the shingle against the roof is 2 lbs. cos 20°, and as the coefficient of friction is 0.5 the total frictional force at incipient motion is 0.5 x 2 cos 20° or 0.9397 lbs. The force causing sliding is 2 lbs. sin 20° or 0.6840 lbs.

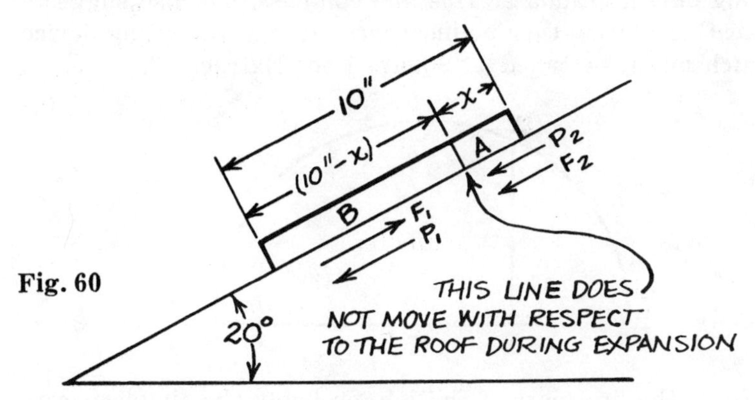

Fig. 60

When thermal expansion causes motion with respect to the roof, the total frictional resistance of 0.9397 lbs. must be developed, yet in general the shingle will not move so that the overall force is 0.6840 lbs. These conditions are met if the difference .9397 — .6840 = .2557 occurs half in one direction and half in the other. Thus a frictional resistance tending to shove the shingle down the roof of 0.12785 lbs. is developed on the heating cycle by the part of the shingle above the center of expansion. This is opposed by a frictional force of (.6840 + .12785) = .81185 lbs., developed on heating, by the part of the shingle below the center of expansion, tending to hold the shingle in place. The center of expansion is $\frac{.12785}{.9397} \times 10 = 1.361''$ below the top of the shingle. On expansion the center of the shingle moves down (5.0 — 1.361) $\times\ 6 \times 10^{-6} \times 50 = .00109''$.

"On contraction, the center of contraction is 1.361″ from the lower end of the shingle and the center creeps down an additional .00109″ making the total movement per day .00218″."

6. SQUARE ROOT EXTRACTOR

This problem goes beyond the normal run, since it not only asks for a way to get the square root of a number by using only a graduated rule and compass, but also suggests that the solution then be incorporated in a computing device which might be called a "Square Root Extractor".

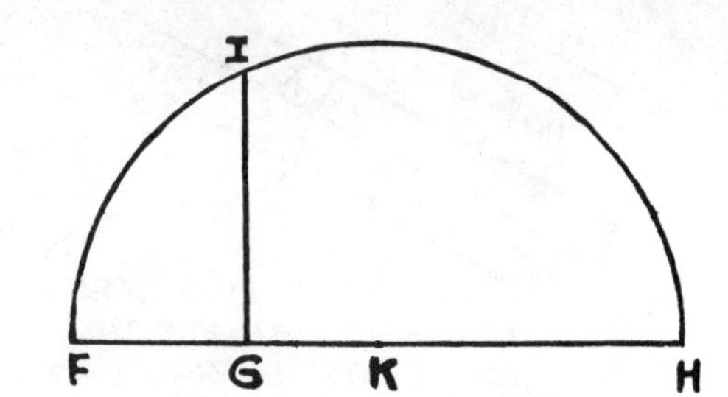

Fig. 61

To do the first part of the job, probably the simplest procedure is to follow the path of the immortal Descartes. About 325 years ago he wrote in his text "La Geometrie", forerunner of analytical geometry, "If the square root of GH is desired (see Fig. 61), I add, along the same straight line, FG equal to unity, then, bisecting FH at K, I describe the arc FIH about K as center, and draw from G a perpendicular and extend it to I, and GI is the required root. I do not speak here of cube roots, or other roots, since I shall speak more conveniently of them later." The explanation, by familiar geometry, is that by this method a right triangle FIH is constructed, of which the altitude GI is the mean proportional to the two segments of the base; one of which is unity and the other the given length KH, whereby $\frac{FG}{GI} = \frac{GI}{GH}$, GI thus being the square root of the given length.

Unfortunately this method does not lend itself too well to incorporation in an extractor, but the same principle was applied to a square root device by B. A. Noble, of Owens-Illinois Glass Co., Oakland, Calif., who dispensed with the circle for producing the right angle and instead used an ordinary right triangle. His sketch shows two scales, AB and OC, perpendicular to each other, with an index A for use with numbers with odd figures before the decimal, and index B for numbers with even figures. The "indicator" for this device is a draftsman's transparent right triangle, one leg

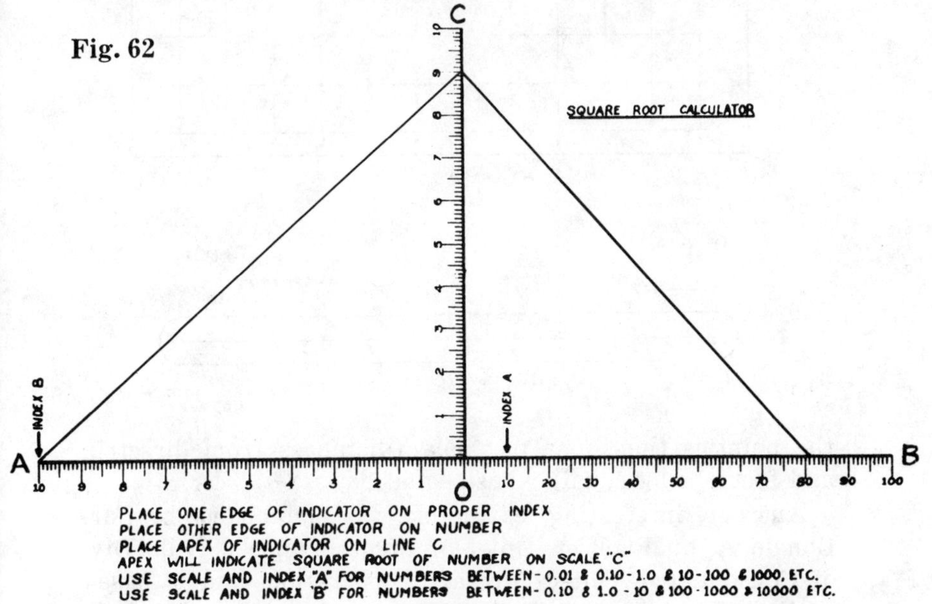

Fig. 62

of which — as shown in Fig. 62 — is placed on the proper index, with the other leg on the number, and the vertex on line OC, which will be at the required square root. It may be noted that closer readings would be had if, instead of the two indexes, Mr. Noble had used just the single index B and two scales on either side of OC, as in other extractors to be described.

Mr. T. R. Mack submitted an ingenious extractor based on the properties of the parabola (see Fig. 63). A stationary scale is cross-ruled like an ordinary sheet of graph paper and on the lower edge (X axis), a point is marked off a quarter of a scale-length to the right of origin 0. At this point is pivoted a rule with scale divisions the same as on the fixed scale. To find the square root of any number as marked by a point on the pivoted rule, the rule is merely swung counterclockwise until the point has an abscissa equal to the number; its ordinate then being the square root. This is because

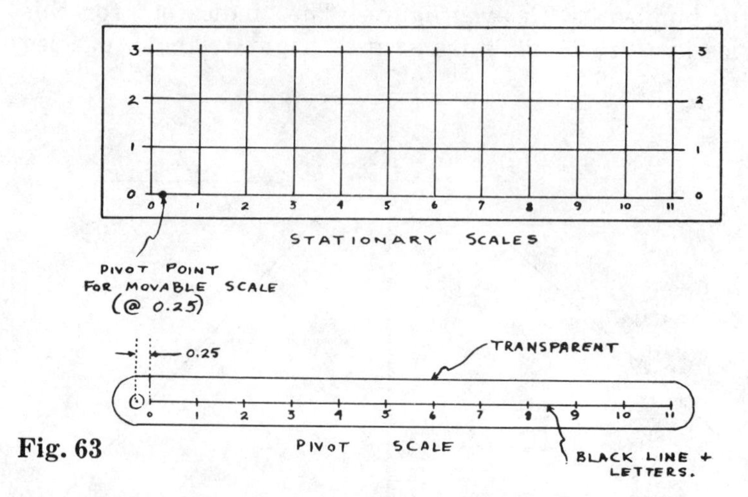

Fig. 63

the point is then on a parabola (distances from directrix and focus being equal) whose equation is $Y^2 = 4AX$.

Another interesting extractor was submitted by Mr. Donald A. du Plantier, Chief of Structures, Nashville Div. of Consolidated Vultee Aircraft Corporation, and is based on the relation that if the hypotenuse of a right triangle is $n + 1$ and one leg is $n - 1$, the other leg will be twice the

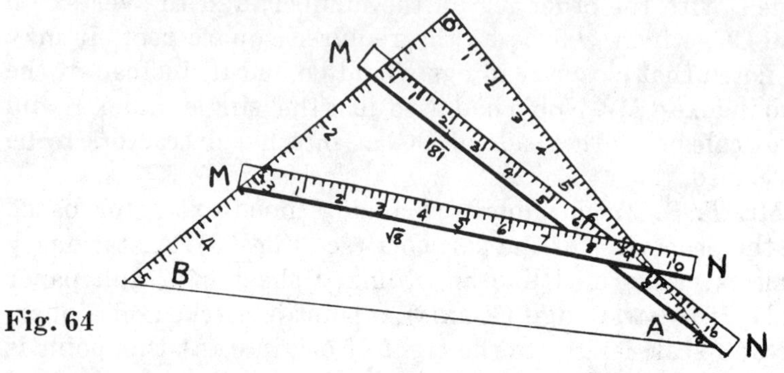

Fig. 64

square root of n (see Fig. 64). It will be noted that the "director" is a draftsmans' right triangle with the graduations on leg OB spaced twice as far apart as on leg OA. The scale MN on which the required number is marked off has the same graduations as on OA. As applied in the diagram to find the square root of 8 (or any number starting with 8 and having an odd number of figures before the decimal point) 7 on scale OA crosses the MN scale at 9 and the OB scale reads the answer 2.83 where it crosses M. The diagram also shows the director applied to find the square root of 81.

Before tooling up for production of any of these devices, the editor offered for consideration the rig illustrated in Fig. 65, based on the relation, not much different from the one used by Descartes, that in a right triangle with altitude dropped to the hypotenuse, either leg is a mean proportional between the hypotenuse and the adjacent segment thereof.

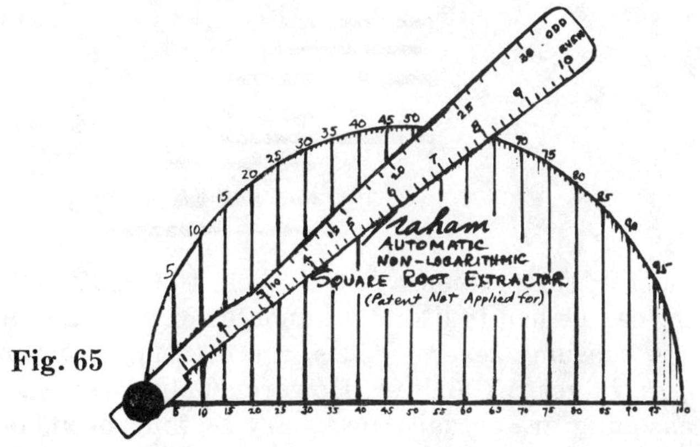

Fig. 65

This avoids the need for indices, since the required square root is merely read on the pivoted scale at the point where it intersects the circle at an abscissa corresponding to the given number. The upper or lower scale is selected depending on the number of figures before the decimal (or the zeros after it, for a fractional number) and it is noted that in the sketch the upper scale reads 23 as the square root of 529, while the lower scale reads 8 as the square root of 64. Incidentally, the graduations to the left of 3.1 on the lower scale are superfluous and not used.

This latter extractor, as in Descartes' method, uses the semi-circle to automatically provide the right angle at the vertex and affords quite close readings. However, no utility other than the relaxation so afforded is claimed for any of the above methods of non-logarithmic extraction, although they do illustrate interesting applications of geometry, both plain and analytical, to invention.

7. THE CHARRED WILL MYSTERY

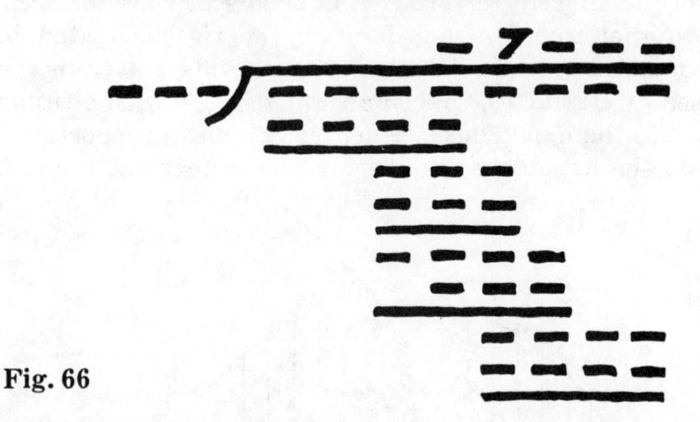

Fig. 66

Like most of the problems in this book, the solution to this mystery can be had in literally hundreds of ways and it is possible to consume several pages, representing a dozen or more steps, to get the answer. However, only three easy steps of reasoning are actually necessary to fully decipher the division, as follows:

1) The fourth figure in the quotient is obviously zero, since two figures of the dividend must be brought down.
2) The first and last figures of the quotient are both bigger than the third because they give a four-figure product, and the third figure in turn is larger than 7, since the third multiplication left a smaller remainder when subtracted from a larger number than did the multiplication of 7. This means that the first and last figures in the quotient are 9 and the third figure is 8; the quotient being thus fully identified in short order as 97809.

3) Since the divisor times 8 is no greater than 999, which is the maximum for the third multiplication, the divisor cannot be bigger than 124. And since the first two figures of the last subtraction can then not be greater than 12, the subtraction being from a four digit figure than can not be less than 1000, the third product must be at least 988, which means that the divisor likewise cannot be less than 124. So there it is! Divisor 124, quotient 97809, which means that the old millionaire left $12, 128, 316 to be divided among 124 heirs.

Now, Mr. Reader, compare the above with your own analysis for brevity and see if the reasoning might be further shortened.

8. FROM POLE TO POLE

This is one of a series of imaginary adventures of little Euclid, the hero of several of the problems included in this volume. Euclid was presumably a prototype of the Quiz Kids, with an instinctive, pristine feel for things mathematical. His precocity proved a bit annoying to some of our readers and over a period of time provoked them to the petulant outbursts that lend color to some of the answers here quoted.

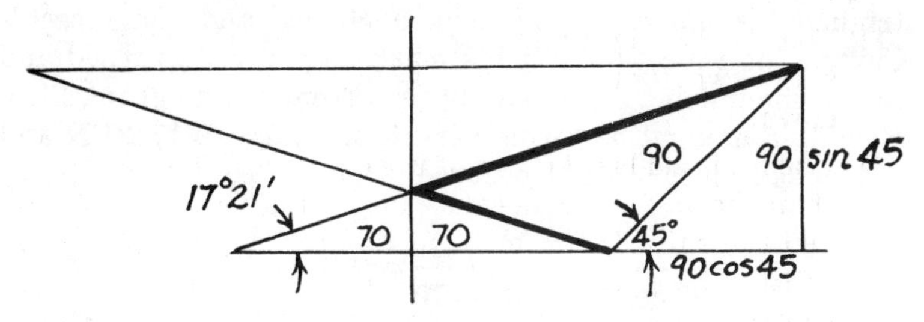

Fig. 67

As in many problems in this book, a solution using the "higher" branches of mathematics, trigonometry and calculus, seems to be the natural mode of attack. Although it provides the correct solution, it is actually much longer and more roundabout than the optimum method. The arduous method is to set up a trigonometric equation (see below) and find the unknown side or angle by setting the first derivative equal to zero. The short cut, of course, is in realizing that no matter what point on the wall Euclid chose, the distance from that point to the starting pole would be the same as if the starting pole had been an equal distance on the other side of the wall (because of the two equal triangles). Clearly, then, Euclid's path would be the shortest if he chose a point on the straight line between such an imaginary starting pole and the finishing pole (as in Fig. 67) since, as little Euclid's namesake taught nearly twenty-three centuries ago, "A straight line is the shortest distance between two points." The word "imaginary" is used advisedly since such a pole would be the "image" of the starting pole if the wall had been a mirror. Obviously, the same analysis would apply to the straight line joining the starting pole with the image of the finishing pole, which inspired E. H. Herbert, of Oliver United Filters, to the quite plausible assumption that the wall was the side of the schoolhouse and had a window which served Euclid better than a slide rule. Wrote Mr. Herbert: "Little Euclid got a break. There was a window in the wall and as he stood at the starting pole he saw the reflection of the finishing pole in the window. Remembering something about the straight line that intrigued his namesake, he took a chance and ran directly toward the reflection, put his mark on the wall, turned and ran to the finish." As is readily seen from the diagram, little Euclid, in navigation terms, struck on course W17°21′N and rebounded on E17°21′N because

$$\frac{90 \sin 45}{90 \cos 45 + 70 + 70} = \tan 17°21'$$

Compare this neat approach with the cumbersome one in five steps by the calculus, which we include by way of exhibit of the correct but circuitous: "Locate C along wall (see Fig. 68) so that (a + b) will be a minimum quantity. RP = TQ.

If we let CQ = x feet then TC = 63.63 — x feet. We now determine for what value of x will (a + b) be the least amount.

1. $a = \sqrt{70^2 + (63.63 - x)^2}$

and $b = \sqrt{x^2 + 133.63^2}$

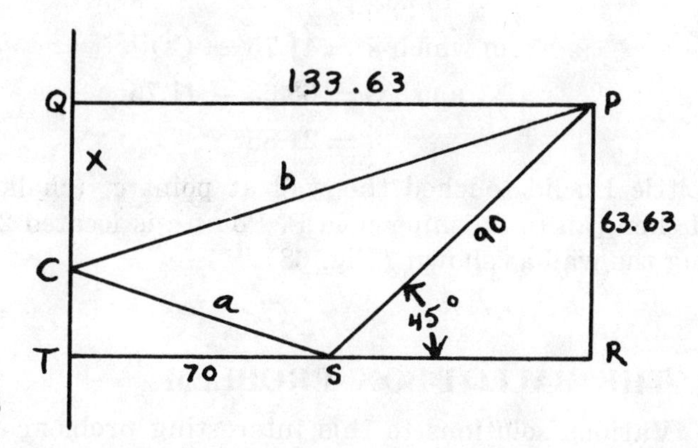

Fig. 68

2. In order to obtain a minimum value of (a + b), we must differentiate the sum with respect to x, and equate the differential to zero to obtain the critical value of x

$$\frac{d(a+b)}{dx} = \frac{da}{dx} + \frac{db}{dx}$$

$$= 1/2\left(\frac{2x - 127.26}{\sqrt{x^2 - 172.26x + 8950}}\right) + 1/2\left(\frac{2x}{\sqrt{x^2 + 17857}}\right)$$

$$= \frac{x - 63.63}{a} + \frac{x}{b}$$

3. Equate $\frac{x}{b} + \frac{x - 63.63}{a} = 0$

$$\text{or } \frac{x}{b} = \frac{63.63 - x}{a}$$

4. This proportion holds only for that value of x which will make Δ STC similar to Δ PCQ.

5. Since ST & PQ are known quantities (70′ & 133.63′) CT & CQ can be found by proportion, or

$$\frac{70}{133.63} = \frac{63.63 - x}{x}$$

$$\text{From which } x = 41.75' = CQ$$

$$\text{and } TC = 63.63 - 41.75$$

$$= 21.88'$$

Little Euclid touched the wall at point c, (chalk mark) and then ran to finishing pole P. Point c is located 21.88 ft. along the wall as shown (Fig. 68)."

9. THE BALLOT-BOX PROBLEM

Various solutions to this interesting problem are possible, using theorems commonly employed in probability. However, the contributor of the problem showed how the thing could be "done with mirrors", quite simply. If A is a vote for winner Anderson and B for loser Brown, then any tally of all the votes cast comprises a "sequence" in some order or other of 300 A's and 200 B's. The required probability is the number of sequences that include a tie at least once in the count, divided by the total number of possible, different sequences. A short-cut in reasoning to determine that quotient readily, is to consider the situation when a tie first appears in any sequence. Say that happens after 8 votes are counted; the tally up to that point going AABABABB. Then regardless of the order of count of the remaining 492 votes, each possible sequence starting with the 8 votes as tallied above can be matched with another sequence in which the remaining votes are unchanged but in which these 8 votes were recorded in exactly reverse order, BBABABAA, "mirrored" as it were from the other arrangement. This reasoning applies if the first tie vote appears after 2 votes are counted, which is minimum, or 400, which is maximum, or anything between. So we can conclude that just half of all the sequences that include a tie must start with A and the other half with B. But obviously all the sequences that do start with B must reach a tie somewhere since A ultimately overtakes B. So the favorable sequences must total twice the number that start with B. For the figures given, 200/(200 + 300) or 2/5 of all sequences start with B, so that the sequences that include a tie are twice that fraction or 4/5 which is the required probability, and for the general figures a and b, it is 2b/(a + b). A spot-check of this analysis for a total of only 5 votes cast, 3 for A and 2 for B, gives the following 10 possible sequences:

1. AAABB	6. BBAAA
2. AABAB	7. BAABA
3. ABAAB	8. ABABA
4. BAAAB	9. ABBAA
5. AABBA	10. BABAA

Of these ten, the last eight all include a tie, four of the sequences starting with A and the other four with B, and we have arranged the sequences so that the "mirrored pairs" follow one another, the first tie occurring at the second, fourth, second, and second votes respectively. The occurrence of additional later ties in the last two pairs is incidental.

A different approach, conceived by Samuel M. Sherman, Missile & Radar Engrg. RCA, Moorestown, N.J., introduces the interesting notion of "starting-points" in any random sequence including all votes; the sequence being considered as a "cycle" of votes. Mr. Sherman showed that of the (a + b) possible "starting points" in any complete sequence, 2b would result in a tie. For a total of say 9 votes, 5 for A and 4 for B, an A vote is shown in the diagram (Fig. 69) by this method as a line moving diagonally upward and a B vote as a line downward. A tie vote is reached when the vote path crosses the horizontal line through the starting point. If we consider any dip in the graph, representing a B vote, there are two and only two independent starting points from which horizontal lines one cycle long will touch or intersect the graph. One of these is immediately before the B vote because the point immediately to the right is lower, whereas the end of the cycle must be higher. The other starting point associated with each unit dip is found by following a horizontal line to the left from the point immediately after the B vote to the first point of intersection or contact with the

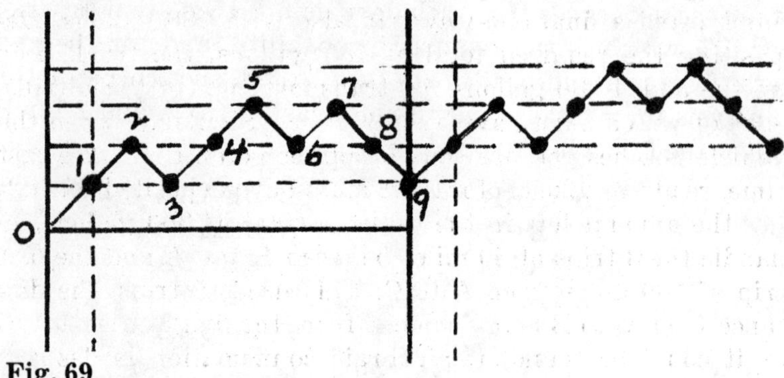

Fig. 69

graph. Thus $2b/(a + b)$ of all sequences gives a tie. Applying Mr. Sherman's analysis to the ten sequences listed above for a 3 to 2 vote, it will be noted that there are only two different cycles: one for sequences 1, 4, 5, 6, 9, and the other for sequences 2, 3, 7, 8, 10. Of the five possible "starting points" in each cycle, four in each result in a tie.

10. TRUCK IN THE DESERT

Many who solved this problem in best fashion questioned the likelihood of such a bizarre occurrence but all agreed that it was a most excellent example of how clear-headed analysis can lead, with least mathematics, to the one correct answer. It helps to diagram the procedure as in Fig. 70, and work backwards from the objective to the starting point. It may be assumed that the figure of 1 mile per gallon is not appreciably affected, for the heavy vehicle, by the amount of gas carried.

Clearly, for best efficiency the vehicle must reach the last point A of dumping with enough gas left in the tanks so that when the amount previously dumped is added to it, the total will be 180 gallons, whereby the last trip can be the maximum distance of 180 miles. Acquiring these 180 gallons calls for a minimum of two forward trips from the previous dump, which means that the vehicle made one round trip starting with a full quota of 180 gallons and returning empty, and a final one-way trip also with 180 gallons. Depositing the required total of 180 gallons, the truck used up the added 180 gallons for transportation in three trips, each of which must have been 60 miles. Similarly, when the vehicle reaches the previous dumping point B for the last time, ready to take off for A, it must have deposited, including the amount left in the tanks, a total of 360 gallons, to handle the 3 trips of 60 miles between B and A and the last trip of 180 miles from A to O. This needs 5 trips, the distance from C to B being evident from the diagram as 180/5 or it could be found, if algebraic confirmation is desired, from the equation $2(180-2X)+(180-X)=360$ from which $X=36$ miles.

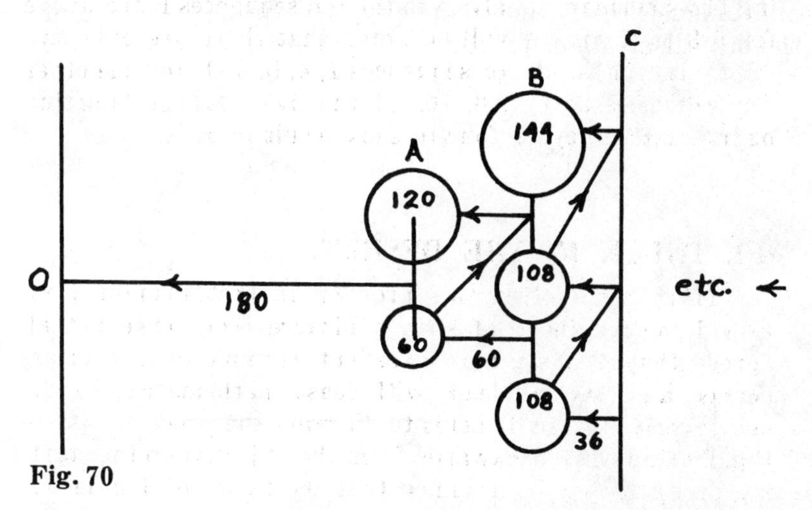

Fig. 70

Clearly the vehicle deposits at each successive dump 180 gallons less than had been accumulated at the previous dump; the 180 gallons being used up in each case for transportation between dumps. The successive distances between dumps, starting from destination, are thus shown to be 180, 180/3, 180/5, 180/7 etc. up to 180/21 or 180, 60, 36, 25.714, 20, 16.363, 13.846, 12, 10.589, 9.474 and 8.571 which add up to 392.557, leaving 7.443 miles (rather than 180/23 which is 7.826 miles) for the most economical distance between the desert edge and the first dump. As explained above, the vehicle must deposit a total of 11 x 180 or 1980 gallons at the first dump, consuming 23 x 7.443 gallons for transportation in so doing, for a grand total of 2151.2 gallons. So that, 180 minus 2 x 7.443 or 165.116 gallons would be deposited on each of the first eleven forward trips and the remainder of 163.724 gallons on the last trip.

11. THE UNRULY BISECTION

Most of the problems in this book were especially selected because they invited widely differing modes of solution, only one of which was outstanding for its unexpected and unique brevity and simplicity. This problem is one of the best illustrations of this challenging quality.

Fig. 71

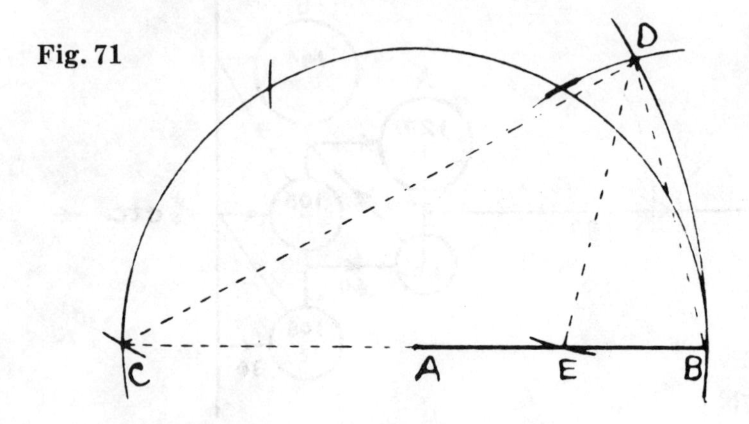

Included in the great number of solutions from readers of the DIAL are many faulty ones in which a straight edge was employed in spite of the injunction against its use. In one case a reader adroitly folded the paper to get a straight line but was "disqualified for folding." Others relied on drawing tangents which were of doubtful accuracy and therefore did not provide a positive solution. However, there was an unusual variety of correctly conceived and proved solutions of varying complexity. Of these the simplest was contributed by R. J. West, of Sears Roebuck, Service and Training, who wrote: "Describe an arc with radius AB and A as center, and describe an arc with the same radius and B as center. With compass still set at radius AB, start from B and strike three successive arcs on the previous arc (center A) to locate point C. With radius CB and C as center, describe an arc cutting the previous arc (center B) at point D. With original radius AB and D as center, strike an arc which will intersect AB at its midpoint E." Note that the broken lines in Mr. West's drawing (Fig. 71) were used in the proof only, which is: "Point C is on the extension of line BA because three successive radii subtend a semi-circle, and triangles BCD and EDB are similar because they are isoceles with a common base angle. Since CD is twice BD, EB must then be half BD, or half AB."

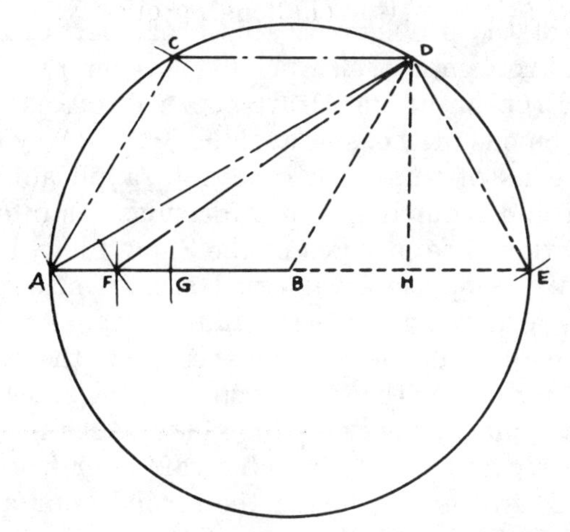

Fig. 72

We include an interesting variation of this solution, illustrated in Fig. 72 and submitted by a DIAL reader, which is just different enough to be wrong. The large circle is drawn as before and the three radii laid off AC, CD and DE. However, for the next step, with E as center and AD as radius, arc F is scribed on line AB, and then with E as center and radius DF arc G is scribed with cuts AB at the supposed midpoint. To find the error in this solution which by actual measurement seems to be right, constitutes an interesting problem in itself; one which the reader of this book may now want to tackle. If you stop at this point and try it as a supplementary problem, you will find, by drawing the lines in the diagram, Fig. 72, and calling $AB = BE = r$, that $BDE = 60°$ because $BD = DE = BE$, from which angle $ADB = 30°$ and angle DAB also equals 30°. $AD = EF = r\sqrt{3}$ from which $FB = r(\sqrt{3} - 1)$ and by dropping the perpendicular DH we have $FH = r(\sqrt{3} - 1/2)$ and since $DH = r\sqrt{3}/2$, the hypotenuse FD which by construction equals EG must be $r\sqrt{(\sqrt{3} - 1/2)^2 + (\sqrt{3}/2)^2} = r\sqrt{3 - \sqrt{3} + 1/4 + 3/4} = r\sqrt{4 - \sqrt{3}} = r\sqrt{2.27}$ instead of $r\sqrt{2.25}$, which reveals the small extent of the error.

12. START OF THE SNOW

Several of the problems in this book, particularly the Pole to Pole Problem, No. 8, Euclid's Put-Put Puzzler, No. 28, and the Problem of the Draftsman's Parabola, No. 29, illustrate ingenious methods of avoiding calculus by employing other means of approach especially applicable to the case in point and requiring no pat formulas for integration or differentiation. The problem of the Snow Plow, however, is one of just the opposite variety. Here, all attacks other than calculus seem to lead to error and fallacy.

Nearly one half of the solutions reached the incorrect figure of 11:30 A.M. for the time of start of the snowfall. We quote from a typical answer of this kind: "Assume that the plow had a removal capacity in cubic yards per hour. If the plow traveled half as far during the second hour as it did during the first hour, the average depth of the snow must have been twice as great during the second hour as it was during the first hour. With snow falling at a constant rate, its average depth may be measured at the half hour. With the plow starting at 12 noon, the depth of snow was twice as great at 1:30 P.M. as it was at 12:30 P.M. Therefore, the snow started to fall at 11:30 P.M.

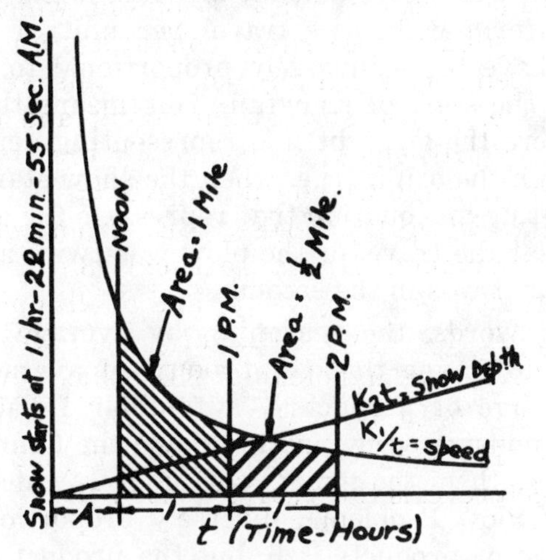

Fig. 73

Others cited the straight line relation of snowfall vs. time to support their error. One especially sanguine contestant wrote as an accompaniment to his erroneous answer, "Simeon Poisson's family tried to make him everything from a surgeon to a lawyer; the last on the theory that he was fit for nothing better. One or two of these professions he tackled with singular ineptitude but at last he found his metier. It was on a journey that someone posed a problem similar to the one given in your January Graham DIAL. Solving it immediately, he realized his true calling and thereafter devoted himself to mathematics, becoming one of the greatest mathematicians of the nineteenth century". On receipt of this letter the author was sorely tempted to concoct a pleasantry turned around the word "Poisson", but refrained.

Actually, the solution by averages is fallacious. Although the depth of the snow does bear a straight line relationship to time and can therefore be "averaged" with respect to time, it is incorrect to assume that the average depth during the second hour must be twice what it was during the first hour, merely because the plow traveled only half as far during the second hour. On the obvious assumption that the plow removed an equal amount of snow in cu. ft. per hour over a uniform width, its travel per unit of time would obviously have to be inversely proportional to the elapsed time from the start of snowfall. This means that in order to find where the straight line representing depth of snow goes through the origin (i.e. when the snow started to fall) we must set up an equation that represents the actual story, which is that the travel of the plow was twice as far in the first hour as it was in the second.

In other words, the reasoning by average depths was equivalent to saying that the product of averages is equal to the average of products — a familiar fallacy which is quickly demonstrated by writing down the figures 1, 2, and 3, and below them the figures 6, 3, and 2, which are, as in the "Snow Plow Problem", inversely proportional thereto. The average of products is 6, but the product of averages is 7-1/3.

Try as we may, this apparently simple problem seems incapable of solution without falling back on Messrs. Newton and/or Leibnitz (history has not decided which of the two merits the discovery of the Calculus) and expressing linear travel of the plow as a summation of momentary rates of travel multiplied by time; the summation during the first hour being twice that during the second. (The actual distance in miles does not apply, merely the ratio.)

Since the volume of snow removed in any instant of time is assumed constant, the product of momentary depth times linear travel must be constant, which means that travel during any instant is inversely proportional to depth. But depth is directly proportional to total time elapsed since the snow falls at constant rate. This gives the equation

$$\int_{A}^{A+1} \frac{k}{t}\,dt = 2\int_{A+1}^{A+2} \frac{k}{t}\,dt$$

where A is duration of snowfall in hours before noon. This reduces to

$$\log \frac{A+1}{A} = 2 \log \frac{A+2}{A+1}$$

$$\text{from which } \frac{A+1}{A} = \left(\frac{A+2}{A+1}\right)^2$$

$$\text{or } A^2 + A = 1$$

positive root of which is $A = .618$, so that the snow started to fall just before 11:23 A.M.

13. THE ROOKIE ELECTRICIAN

This interesting problem, although it calls upon no branch of formal mathematics for its solution, is a type commonly included in supplementary mathematical texts because it illustrates typical mathematical reasoning and procedure in striking fashion. Mathematics has been characterized as "a method making for the economy of thought" — and certainly the one simplest procedure to follow in solving this problem calls for utmost clarity of thought and economy of effort by the supposed "rookie". So much so that after the optimum solution as described below appeared in the DIAL, one of our readers, Albert R. Kall, Consulting Engineer, Ark Engineering Co., Philadelphia, took pen in hand to tell us he was intrigued by the spectacle of (we quote) "the rookie electrician who, in the course of one brief problem, is first unwilling (or totally unable) to perform the original task ordered by his foreman, and then, upon reporting his failure, is ordered nevertheless by the foreman to do the job in the most efficient manner. What happens? Of course the now no-longer rookie electrician does the job, perhaps under divine guidance, with any one of several ingenious solutions. The one presented would do justice to an engineer or mathematician of the first rank, one far more skilled than our formerly dumb electrician. I think such complete and nearly instantaneous improvement in the ability of our problem electrician is exemplary and deserving of some sort of award."

As a matter of fact, the solution submitted by James I. Leabman, AEEL Radar Division, Naval Air Dev. Center, Johnsville, Pa., impressed us as a gem of succinct expression. We quote it verbatim and, since the method is applicable to any odd number of wires, we have, for simplicity, illustrated the three steps in the diagram using only 11 wires. Wrote Mr. Leabman: "Rookie groups twenty wires into 10 pairs, twisting each pair. This leaves one wire by itself. He shorts each twisted pair and then rows across river, taking an ohmmeter or buzzer with him. He locates each twisted pair by a continuity check of all the wires. He labels the pairs 1a, 1b; 2a, 2b; 3a, 3b; . . . 10a, 10b. The odd wire is the 21st which he labels G. He shorts G to 1a, 1b to 2a, 2b to 3a, 3b to 4a, . . . 9b to 10a. He rows back and removes the shorts but leaves the pairs twisted. He then checks continuity between G and the other wires, locating 1a which he labels as such. 1b is then obviously 1a's twisted mate. He then checks continuity between 1b and the other wires, locates 2a and labels it. 2b will be 2a's twisted mate. He follows this procedure until he locates 10a, thereby finding 10b as 10a's mate. He has thus identified all the wires."

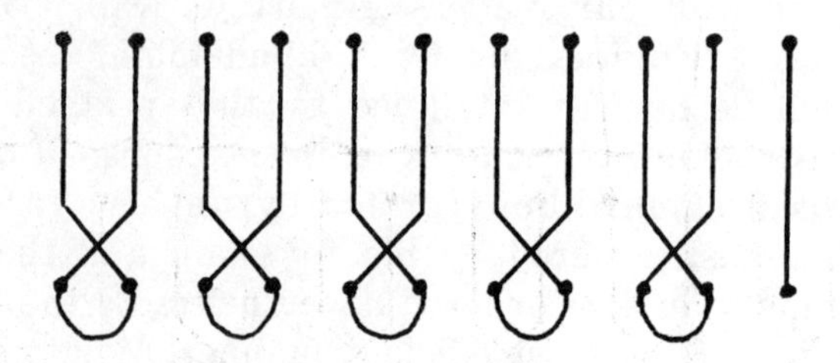

Fig. 74

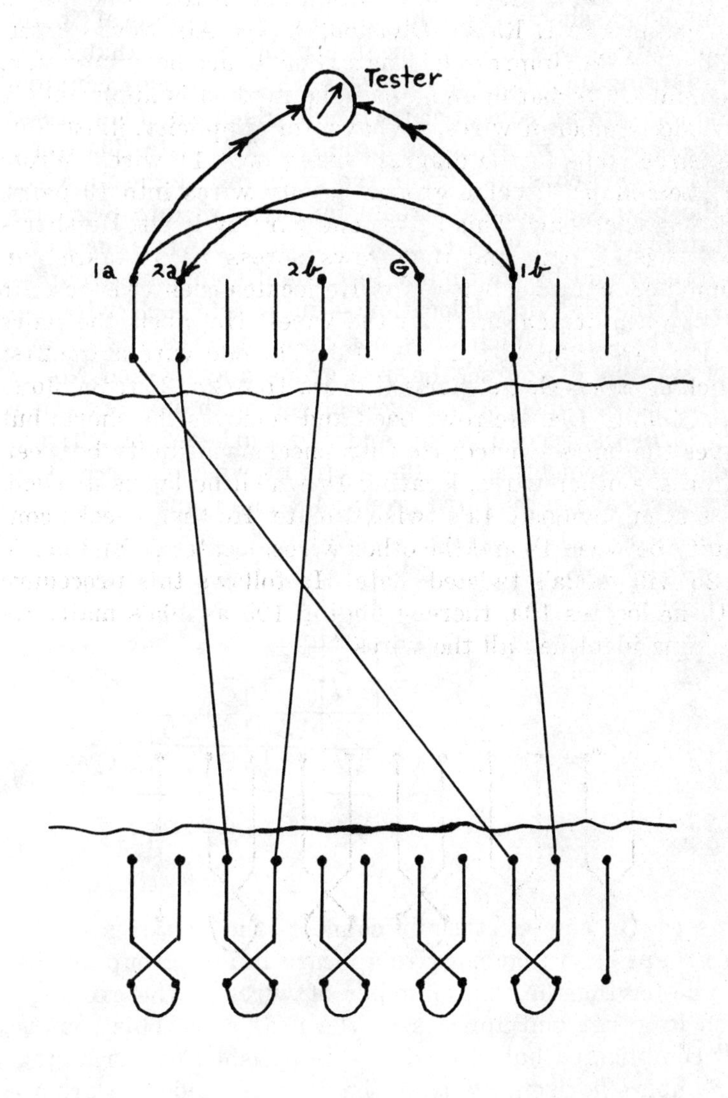
Tester
1a
2a
2b
G
1b

By way of contrast we include an alternative procedure submitted by many readers which likewise calls for only one round trip by the rookie but adds greatly to the complication. This scheme is applicable to any "equilateral" number of wires, by which we mean a number such that if we were talking of coins (as in problem No. 70) they could be made to form an equilateral triangle. Such numbers are of course 6, 10, 15, 21 as in this problem, 28, 36 etc. The principle involved consists first in the grouping of the 21 wires on the near side of the stream by jumpers into 6 groups containing respectively 1, 2, 3, 4, 5, and 6 wires, crossing the stream and by continuity test, identifying the group — but not the individual location — in which each wire belongs. The wire which does not make a circuit with any other wire is, of course, the first group and marked No. 1; the two wires which each make a circuit with only one other wire form the second group and are marked 2 and 3; the three wires which each make a circuit with either of two other wires are in the third group; 4, 5 and 6 and so on.

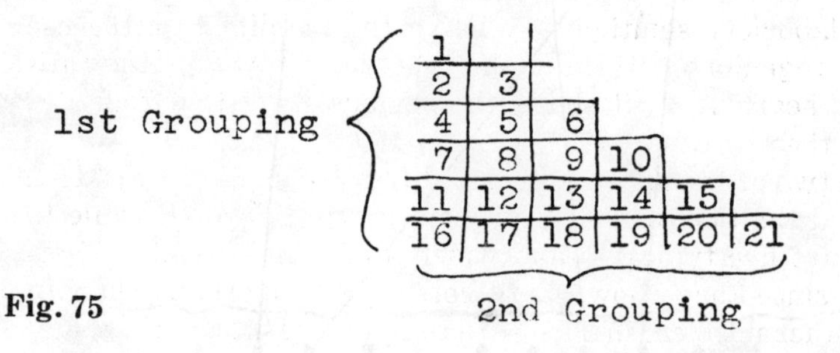

Fig. 75

The rookie now electrically connects the 21 wires into 6 new groups such that no wire appears in any group on the far side having the same number of wires as the group in which its other end appears on the near side. This sounds a bit complicated but is readily accomplished by arranging the numbers horizontally as in the diagram and then grouping the vertical numbers; the purpose is accomplished even if the numbers in any group are not arranged consecutively.

Rookie now rows back, disconnects the original groups, after marking the wires as to the group they were in. Then with the continuity checker he sorts them into the 6 groups as connected on the other side. Now the wires can be numbered because each has a unique position in the two groupings. For instance the wire which was originally connected in the horizontal group of 5 and winds up connected in the vertical group of 4 must be the wire marked No. 13 on the far side, and can be so labeled.

Certainly, this method does not compare in simplicity with the pairing method which, though described for an odd number of wires, can also be used with a slight modification for an even number.

14. THE POOL TABLE

This relatively simple problem in trigonometry is nonetheless of special interest since it illustrates how an otherwise laborious solution may be vastly simplified by the use of an ingenious artifice of the type dear to the mathematician's heart. Also, it shows the devious ways that folk find to do the self same job.

No two of the answers received from readers of the DIAL tackled the task in the same way. It will be worth while to compare five typical — and widely different — modes of attack, since they show, more convincingly perhaps than in any other problem in this collection, how the true mathematician finds ways to lighten labor.

First, the direct attack by straight algebra, applied to the well-known theorem of Pythagoras. Perpendiculars were dropped as shown in Fig. 76 by a reader who evidently had plenty of time on his hands, and three simultaneous equations set up as below.

(1) $x^2 + 9.5^2 = z^2$ or $x^2 + 90.25 = z^2$

(2) $y^2 + 11.5^2 = z^2$ $\quad y^2 + 132.25 = z^2$

(3) $(x + y)^2 + 2^2 = z^2$ $\quad x^2 + 2xy + y^2 + 4 = z^2$

The algebra then follows its unerring course, leading finally to an equation of the fourth power.

$$3x^4 + 88.5x^2 - 16{,}448 = 0$$

from which $x = 7.794''$
$y = 4.33''$
$z = 12.29''$

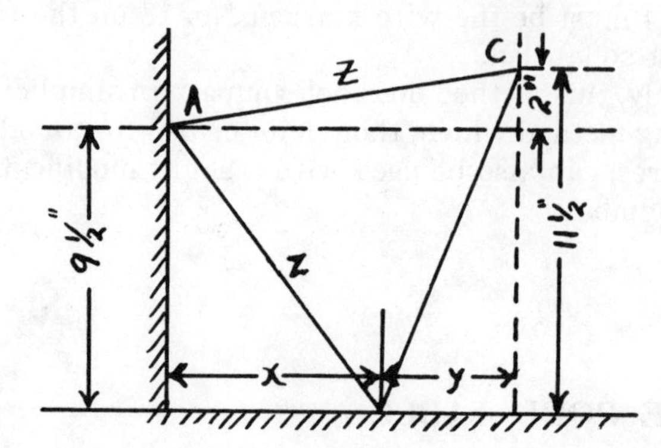

Fig. 76

Second, the approach by similar triangles, as shown in Fig. 77. Here four construction lines are used, the median to side AC and three perpendiculars as indicated. This gives the two similar triangles BED and ACG (their sides are respectively perpendicular) and the relation CG/AC = BE/BD. But BD = AC cos 30° and BE is obviously equal to 11.5 less half of 9.5, or 6.75. Substituting these values gives CG = 6.75/cos 30° or 7.794″.

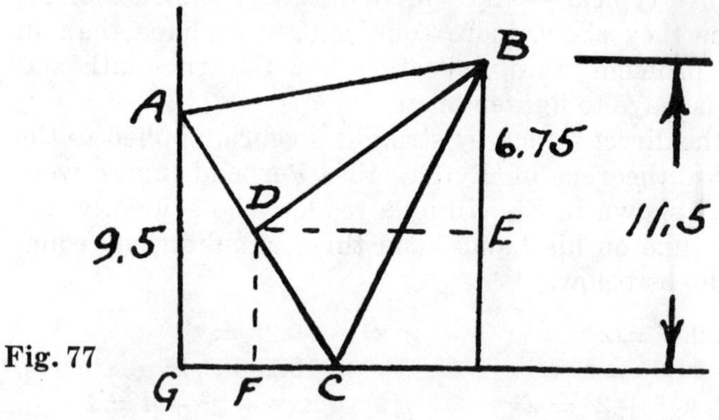

Fig. 77

Third, the approach by subtended arcs. Here the ingenious artifice used is that of describing an arc of a circle with point C as center and passing through B, A, and P.

Then, $PQ = AQ = 11.5'' - 9.5'' = Z''$
Arc AB is 60°, since $\measuredangle C = 60°$
$\therefore \measuredangle BPA = 30°$ (measured by 1/2 arc)
$\therefore PB = 2OB = 2X$
or $3X^2 = \overline{OA + AQ + PQ}^2 = 13.5^2$
$X = 13.5 \sqrt{3}/3 = 7.794$

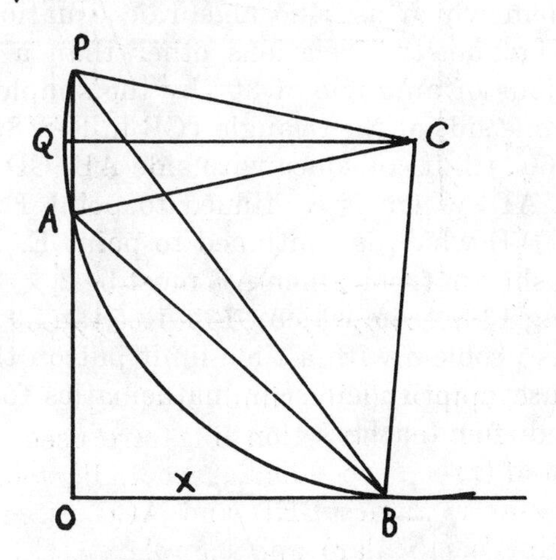

Fig. 78

Fourth, the approach by trigonometric formula, using the artifice of the equation for the sine of the sum of two angles.

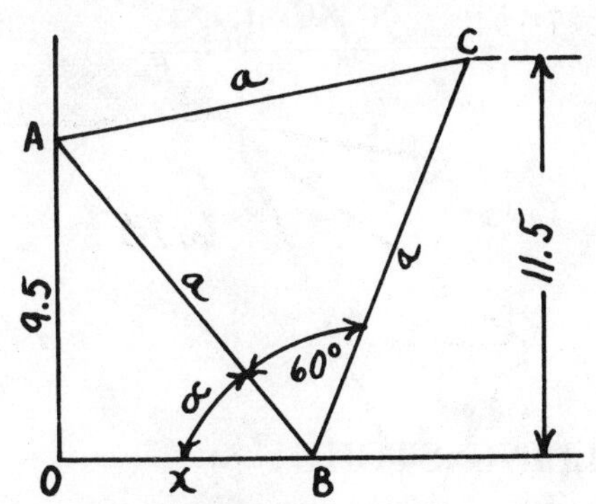

Fig. 79

This involves perhaps the least labor of all. From Fig. 79 it is seen that

$$a \sin \alpha = 9.5$$

$$a \cos \alpha = x$$

$$a \sin (\alpha + 60) = a \ (\sin \alpha \cos\ 60^\circ + \cos \alpha \sin\ 60^\circ) = 11.5$$

$$(9.5 \times 0.5) + 0.866x = 11.5$$

$$\text{whence } x = \frac{6.75}{0.866} = 7.794$$

And fifth and finally, the attack as submitted by the author of the problem which uses no algebraic equations and requires no trigonometric relations other than a knowledge of the functions of an angle of 30° by the simple expedient of swinging one side of the triangle (CB in Fig. 80) through an angle of 60° till it coincides with side AB. CD then takes the position AD′ which is continued to point F. DB takes the position D′B which is continued to point E. Angles are obviously as shown, from which AE must be 2 × 11.5 or 23″, OE then being 13.5, from which OB is 13.5 tan 30° or 7.794″.

Surely this problem with a time limit put on the solution could serve as an aptitude test in mathematics for students, engineers and even teachers.

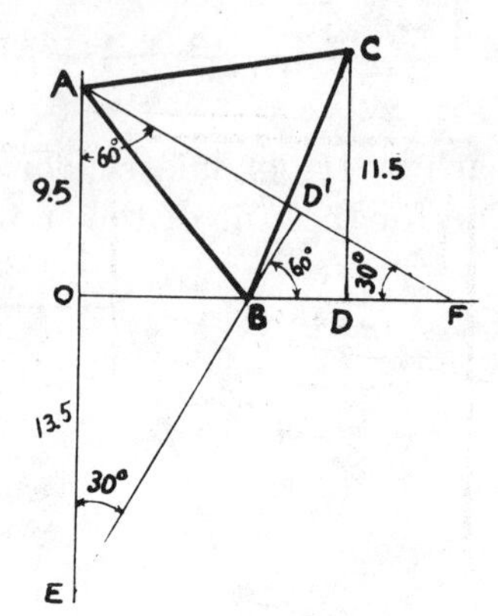

Fig. 80

15. THE FALLING STONE

Since the statement of this particular problem specifically enjoins that the job be done in the shortest possible space — on the back of half a business card or on an air mail stamp — we are reproducing, in actual size, four such solutions of varying merit, and the outstanding solution which was judged best of those received. It will be noted that although many solutions answered the requirements as to space and tackled the problem from various ingenious angles, only one of the great number received sidestepped the calculation of the time of fall, which was not asked for. Of all those that did calculate the time as an intermediate step in the solution, the most interesting was Mr. McKenna's (see Fig. 81), who adroitly stated that if the time to fall the first half of the wall were t seconds, then the time to fall the entire height would have to be $\sqrt{2}t$ seconds, so that $\sqrt{2}t - t$ would then be .5 seconds, from which t is 1.71 seconds, which multiplied by g/2 to give the required answer of 46.6 feet.

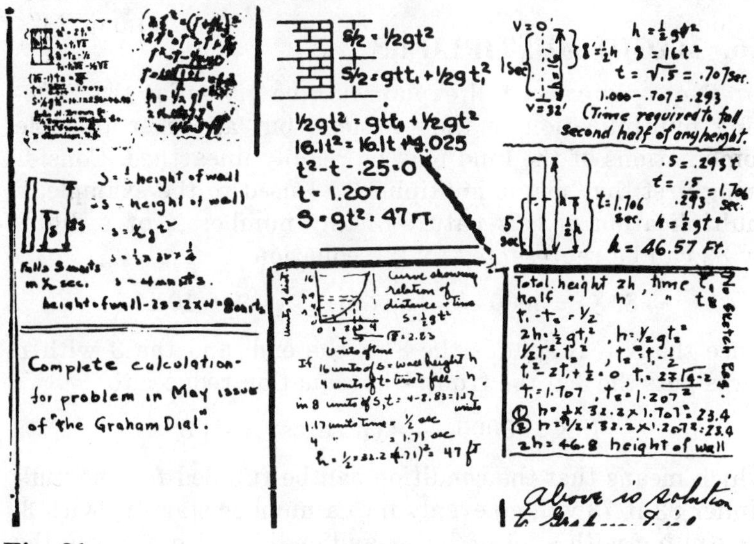

Fig. 81

However, the solution by Charles F. Pinzka, of the American Cyanamid Company, (see Fig. 82) was clearly the best, since it avoided quadratic equations, dual roots, velocities, time of fall, and extraneous elements of any kind. It reached the answer rapidly and unerringly by first establishing from the basic equation $S = 1/2\,gt^2$ that in any fall the square root of the distance fallen, taking g as 32 ft. per second per second equals four times the time of fall. From this the difference between the square roots of the unknown height and half that height must equal one half second times four, which gives the answer through a single, simple equation.

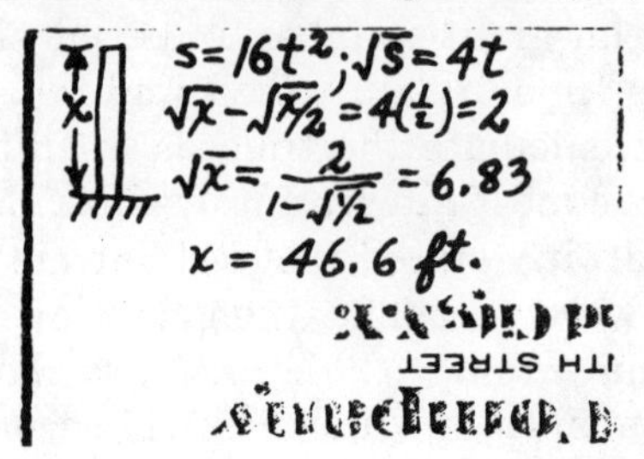

Fig. 82

16. MAGIC MULTIPLIERS

Not only can a full explanation be given for the lightning multiplication in the problem, but all other possible combinations of the kind may be readily unearthed. Considering first the "magic multiplier" 83 used in the example, a multiplication of this nature of any number X of n digits by 83 can be represented by the equation:

$$83X = 3000\text{ - - - (n zeros) - - - }8 + 10X$$

since the 8 + 10X adds the 8 at the end, and the 3 with n zeros puts the 3 at the front. This equation reduces to

$$73X = 3000\text{ - - - (n zeros) - - - }8$$

which means that the condition can be fulfilled for the multiplier 83 if 73 will go evenly into a number starting with 3, continuing with a lot of zeros, and ending with 8. Since the last figure in the quotient must obviously be 6, which multiplies 73 to give 438, this is the equivalent of saying that in the continued division a remainder of 438 must ultimately appear, which of course will not take place if and after any other remainder repeats itself in the process. When we divide 3000 – – – 8 by 73 we get a remainder of 438 after the quotient reaches 41096, which gives us another answer for 83 having 8 figures less than the example given. Actually, there is an infinite number of values for the multiplier 83, each with eight digits more than its predecessor (see below).

Obviously, by checking all two digit multipliers in this way from 10 to 99 every such case could be discovered, but the task would be unnecessarily arduous and the number of eligible multipliers can be vastly reduced by applying four criteria of the Diophantine variety.

To begin with, the second number of the multiplier cannot be 0, so that knocks out 9 possible multipliers at first crack. Secondly, the first figure of the multiplier must be larger than the second, since otherwise the first figure in the product could not be as large as the second figure in the multiplier, which is here required. This now throws out all possible multipliers except 21, 31, 32, 41, 42, 43, 51, 52, 53, 54, 61, 62, 63, 64, 65, 71, 72, 73, 74, 75, 76, 81, 82, 83, 84, 85, 86, 87, 91, 92, 93, 94, 95, 96, 97 and 98,— thirty-six in all. Thirdly, it is clear that some multiplier of the second figure must give the first figure,— and this criterion now discards all but 21, 31, 41, 42, 43, 51, 53, 61, 62, 63, 64, 71, 73, 81, 82, 83, 84, 86, 87, 91, 93, 97,— reducing to twenty-two possibilities. Fourthly, we can now apply the familiar rule of nines and since we know that the "nines" in multiplier and multiplicand must both add and multiply to give the "nines" in the product, this limits us to the combinations 2 and 2, 3 and 6, 5 and 8, and 9 and 9, throwing out 1, 4 and 7 and with them all possible multipliers but 21, 41, 42, 51, 53, 62, 63, 71, 81, 83, 84, 86, 87 and 93,— now only 14 possibilities.

It does not take long to check each of these by the method described above and to find that 71 and 86 are the only other magical multipliers. Eighty-six, strange to say, works with the single figure 8 to give 688 as well as with 7894736842105-263158, and 71 not only pairs with the elephantine number 1639344262295081967213114754098360655737704918032787 but with a group of astronomical numbers made by repeating to the left, as often as desired, the above number with the last figure left off and with 688524590 tacked on at the end, the number of figures in the multiplicand thus being 52, or 112, or 172, etc.

17. THE FOUR FOURS

Major Hitch's problem of the Four Fours was conceived during the height of World War II when the Major was in the U.S. on a special mission for the R.A.F. "It provoked more clamor," we quote from the Private Corner for Mathematicians in the DIAL, Nov. 1943, "than has resounded in this quiet corner for many a moon." In several cases we received two or more letters from the same contestants on successive days or in successive mails. One reader submitted a total of no less than five different solutions, all of them brilliantly conceived. Although the problem was artless enough, many resorted to intriguing devices such as gamma functions, integrals, trigonometric functions, Roman numerals, imaginaries, zero powers, logarithms, weights and measures and the like. Many of them though far-fetched and cumbersome were more interesting than the optimum an-

$$\frac{4\text{ deg. }44\text{ min.}}{4} = 71\text{ min.}$$

$$\frac{4^4 + 4!}{4} - (i \times i) = 71$$

$$\left(\frac{44}{4}\right)\text{deg} + \text{arc sec}\sqrt{4} = 71\text{ deg.}$$

$$\frac{4C}{4} - \frac{C}{4} - 4 = 71$$

$$4\text{ ft.} + \left(4! - \frac{4}{4}\right)\text{in.} = 71\text{ in.}$$

$$4 \times 4! - [\sec(\arctan\sqrt{4})]^4 = 71$$

$$4\text{ lb.} + \left(\frac{4! + 4}{4}\right)\text{oz.} = 71\text{ oz.}$$

$$4 + IV + (4 \times 4) = \frac{4\ IV}{55} + 16 = 71$$

$$\frac{\sqrt[.4]{4} - .\dot{4}}{.\dot{4}} = 71$$

$$\frac{4!\sqrt{4} - \sqrt{.\dot{4}}}{\sqrt{.\dot{4}}} = 71$$

$$\frac{4! + 4.4}{.4} = 71$$

Fig. 83

swer. Several of these are included in Fig. 83. It is significant to note that when the problem was published, Major

Hitch himself and the Editors had no answer but the next to last, which as it turned out was sent in by nobody at all and is quite unwieldy as compared to the answer below it, submitted by quite a sizable group of readers.

18. PROBLEM OF THE EASIEST THROW

This is one of several problems in our collection (see also No. 29, Curvature of the Draftsman's Parabola and No. 6, Square Root Extractor) which illustrate the properties of the parabola, perhaps the most fascinating of all curves. As any reader of this book knows, the path of a projectile is a parabola whose horizontal travel x from the starting point at any instant is equal to the elapsed time t multiplied by a constant m, which is the horizontal component of the initial velocity. The vertical travel y equals the elapsed time t multiplied by the average upward velocity which equals the constant vertical component n of the initial velocity minus at/2, which is the average downward velocity due to gravity a. Since $x = mt$ and $y = nt - \frac{at^2}{2}$ (as immortalized in verse by W. S. Matthews, (Rhyme No. 10 opposite p. 23) the equation of flight by substituting $\frac{x}{m}$ for t becomes $y = \frac{nx}{m} - \frac{a}{2} \cdot \frac{x^2}{m^2}$ which is a parabola. Now the roundabout way to solve this problem is to derive some of the properties of the parabola, whereas the direct approach is merely to select the supposedly familiar property most apt to the case and so put it to use that the problem actually becomes one in mental arithmetic. By way of introduction and convenient review we set down some of these familiar properties, with a word or two as to how they are derived.

The parabola, by definition, is generated by so moving a point that its distance from a fixed point, called the focus F, remains equal to its distance from a fixed line, called the directrix DD (see Fig. 84). The parabola, incidentally, is

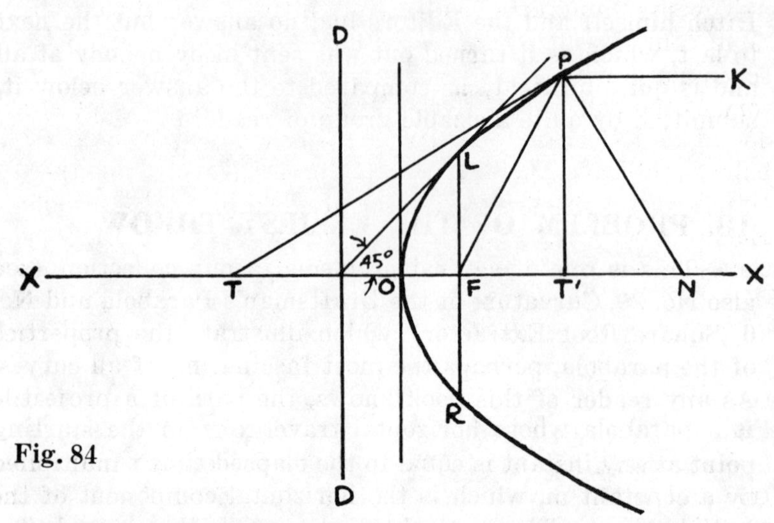

Fig. 84

also the boundary of the conic section formed by cutting the cone with a plane parallel to its axis. The parabola is obviously symmetrical to the line passing through the focus and perpendicular to the directrix, which is called its axis XX; and its origin O, which is the point where the parabola crosses the axis, must by definition, be midway between the focus and the directrix. The chord LR perpendicular to the axis and passing through the focus is called the latus rectum, and its length LF above the axis is of course equal to its distance from the directrix. It forms an angle of 45° with the tangent drawn to the parabola at the point where it intersects the parabola. Because of this it is readily shown in ballistics that for greatest "range" or distance traveled for a given starting velocity, where the firing angle must clearly be 45°, the gun will be on the latus rectum of the parabola of flight. Other familiar and readily proved properties of the parabola illustrated in the diagram are that for any point P the origin bisects the subtangent TT′, the length of the subnormal T′N is constant, and the normal NP at any point makes equal angles FPN and NPK with two lines PF and PK drawn from the point to the focus and parallel to the axis. The latter property is, of course, the one used in parabolic reflectors.

Now assuming little Euclid had all this at his finger tips, he of course realized that to clear the 40 ft. "range" of the schoolhouse with least starting energy the parabolic trajectory of the ball would have to reach and leave the roof at a 45° angle (several readers were in error in saying that the ball should leave Euclid's hand at that angle, the angle there actually being 60°). Knowing the properties of the parabola, Euclid at once realized that its focus would then be at the center of the roof; the roof being the latus rectum, and its directrix half the width of the roof (20 ft.) thereabove, or 40 ft. above Euclid's hand. This in turn means that said hand, being on the parabola, would also be 40 ft. from the roof center. Clearly then, from the right triangle BGF (see Fig. 85) Euclid's distance from the wall was 20 times the difference between the square root of three and unity, or (mentally) 14.64 ft. This solution is probably the neatest since it avoids all calculation of time or velocity.

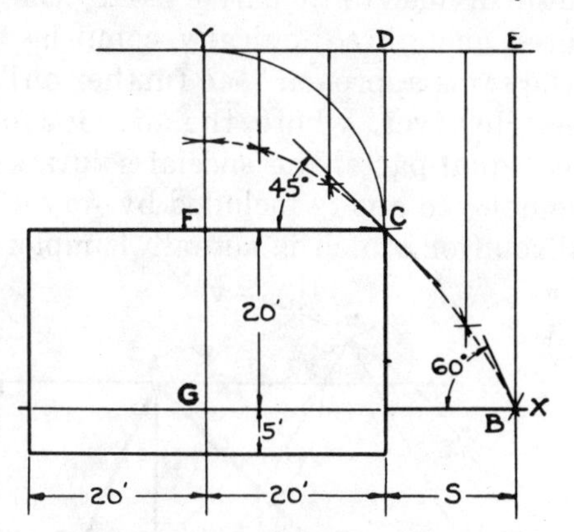

Fig. 85

But one other approach also deserves mention. It puts to use the relation known in ballistics that a projectile to cover a "range" with minimum starting velocity rises to a height equal to one fourth that range. (This is of course the same as saying that the length of the latus rectum is four times its distance from the origin, which is clear from our introductory analysis.) Wrote that reader, "By well established theory in covering a range of 40 feet with minimum starting velocity the ball will rise 10 feet at the top of its flight. It rises 20 feet from point of delivery to edge of wall, or 30 feet total. Calling T_1 the time to rise 30 ft. and T_2 the time to rise the last 10 ft. and since the diminution in vertical travel in a given time arises only from the force of gravity which is governed by the familiar equation $S = 1/2\ at^2$, we have

$$\frac{T_1^2}{T_2^2} = \frac{3}{1} \text{ or } \frac{T_1}{T_2} = \frac{\sqrt{3}}{1}.$$

Horizontal travel during T_2 is 20 feet. So during T_1 -T_2 it is $(\sqrt{3} - 1)$ 20 feet, which is 14.64 feet.

19. THE SLICED TRIANGLE

Although divided into three parts, the questions are closely related enough to logically comprise a single problem, since the same approach (see further on) can be applied to all three. However, where the division of the triangle is into three equal parts, the special solution shown in Fig. 86 may be employed and is included by way of contrast with the general solution which is actually simpler. For this first

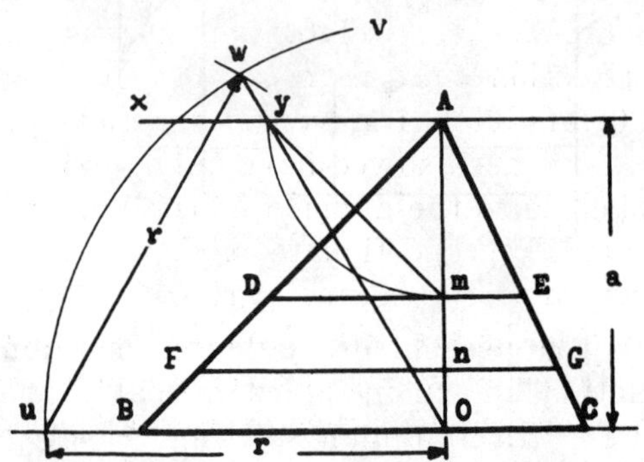

Fig. 86

part, drop perpendicular AO. With O as center and any convenient radius r, swing arc intersecting extension of CB at u, and then with u as center and same radius intersect this arc at w. Join Ow and draw Ax parallel to BC and intersecting Ow at y. Lay off Am on AO equal to Ay, join my and lay off An on AO equal to my. Draw required lines DE and FG parallel to BC. Proof: Since angle AOy is 30°, Am which equals Ay, must equal $AO/\sqrt{3}$ and $An = ym = \sqrt{2AO}/\sqrt{3}$. Since similar triangles have areas proportional to the squares of corresponding sides or altitudes, triangles ADE and AFG are respectively 1/3 and 2/3 area of triangle ABC.

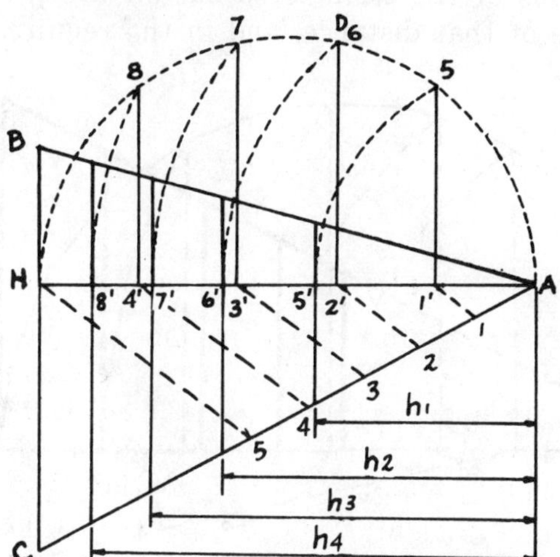

Fig. 87

It will be noted that this solution applies only to the special case of trisection since it makes use of the properties of angles of 30° and 60°. However, as mentioned above, the problem could have been solved in equally simple fashion by a method applicable to the division of the triangle into any number of equal or proportionate parts. In the case of 5 equal parts, which is the second part of the problem (see Fig. 87), equal lengths are laid off on one leg AC and parallels drawn in the usual way to properly divide the altitude AH, on which a semi-circle is then drawn and perpendiculars erected at the division points. Arcs are scribed with A as center, from the intersection of the altitudes with the semicircle to the altitude, and the required lines drawn parallel to the base. The method used is obviously quite similar to that used in the Graham Square Root Extractor.* The dis-

**See* Problem No. 6.

tance from A to the intersection which is swung back to the altitude, is in each case the leg of a right triangle (because it is inscribed in a semi-circle), with altitude dropped to the hypotenuse dividing it into two segments, and is therefore the mean proportional between the whole hypotenuse and its adjacent segment. Its lengths are then successively the square roots of 1/5, 2/5, 3/5 and 4/5 of the altitude, whereby the portions of the triangle so cut off are proportional to the square of that distance, and in the required ratio.

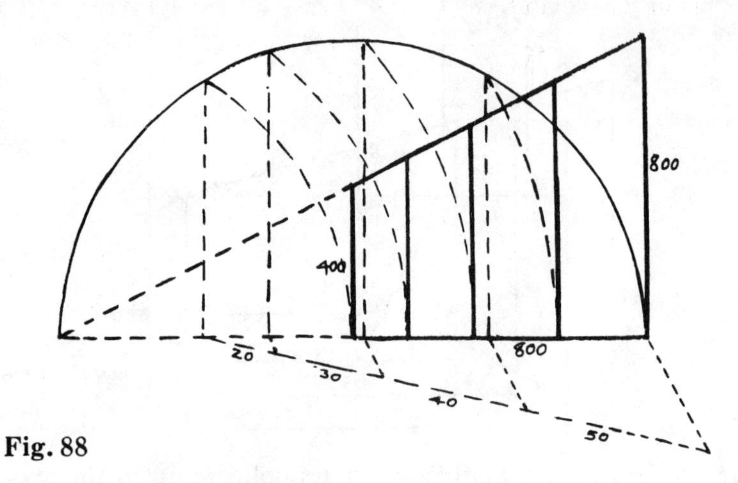

Fig. 88

The third part of the problem merely extends this method in reverse to a trapezoid with the divisions now made unequal. As shown in Fig. 88, the trapezoid is extended to form a triangle, a semi-circle is drawn on the base, an arc swung to the circumference from the point on the base marked by the edge of the trapezoid, and a perpendicular dropped back to the base. This will cut off a fraction of the base obviously equal to that fraction of the trapezoid plus the triangular extension which is represented by the extension. The remainder of the base is then divided into sections corresponding to 20, 30, 40, and 50 respectively; perpendiculars are erected, arcs swung, and parallels drawn as in the preceding part, which completes the job for the reason previously explained.

20. THE CLOCK WEIGHTS

This interesting problem can of course be solved by making an enlarged scale drawing showing the fall of each weight at each time interval and finding the greatest separation by actual measurement. Or the heights before and after striking may be tabulated in a chart, to give the required answer. This approach is doubtless the simplest, but since Mr. Gilbert asked especially for higher treatment, if possible, by calculus or otherwise, the palm goes to Mr. W. W. Johnson, Cleveland engineer, for his ingenious solution by

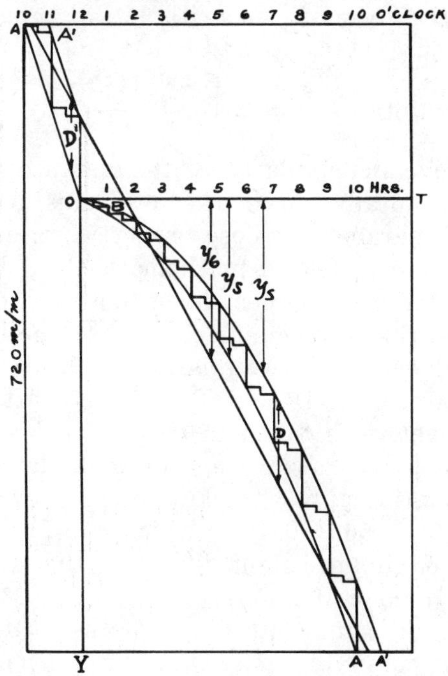

Fig. 89

the calculus. In Mr. Johnson's diagram (Fig. 89) the zero point is taken at 12 o'clock since at this point the striking sequence becomes discontinuous, and the graphs in the top left-hand corner are continuations of those at the bottom right. The path of the time weight, which moves 720/12 or 60 mm./hr., is then the straight line whose equation (ordinates below the axis being considered positive) is $y = 60t - 95$. The path of the striking weight, which makes a total of $12(1 + 12)/2 + 12 = 90$ strokes or 8 mm. plus stroke, though actually the serrated line, can be represented at the even hours by the parabolas in the diagram. Their equations are found to be $Y = 4t(t + 1)$ before striking and $Y = 4t(t + 3)$ after striking, remembering that only integral values of t may be used. This is because just before striking each hour t the striking weight has made t previous half strokes and $t(t - 1)/2$ hour strokes; the latter figure being found from the familiar equation $\frac{n}{2}(a + 1)$ for the sum of the terms in an arithmetical progression, n being equal to $(t - 1)$, $a = 1$ and $L = t - 1$. In terms of distance fallen the total is $y = 8\left[t + \frac{t(t-1)}{2}\right] = 4t(t + 1)$ as given above and similarly for the distance fallen after striking. (Incidentally, three other problems, Nos. 6, 18, and 29, deal with parabolas, whose properties offer fascinating adventures in the field of mathematics.)

The difference between the ordinates of each of the parabolas, and that of the straight line of the time-weight path is then given by the two relations $D = 4t^2 - 56t + 95$ (before striking) and $D = 4t^2 - 48t + 95$ (after striking). The first derivative is taken of each of these expressions and examined in the usual way for maxima and minima; the upper equation giving the key to the answer from the relationship $8t - 56 = 0$, from which $t = 7$ (an integral value) and the maximum height difference then equals $(4 \times 7^2) - (56 \times 7) + 95$ or 101 mm.

Our readers may question whether this adequately excludes the possibility of a maximum at a half-hour point, but a little consideration will show that since the striking weight drops 8 mm. at the half hour whereas the time weight drops 30 mm. during a half hour, the greatest separation must come at the even hour. If the striking weight were above the clock weight at the half hour it would be further from it at the next even hour, and if it were below it at the half hour it would have been further away at the previous even hour. Q.E.D.

21. THE MISSING BEECH

When this problem appeared in the DIAL, the lost tree failed to "stump" a good sprinkling of readers who not only gave a perfect solution but in many cases wrote imaginary sequels to the episode, which ranged from gay to gory.

Unfortunately, many who located the treasure properly did not give their reasons, and others based their answer on the *a priori* assumption that the beech wasn't necessary (merely because we had made a problem out of it) and found the treasure by taking limiting cases of the beech's location, such as at the apex of an isosceles triangle or other possible points. The flaw in that approach is that little Euclid himself presumably didn't know that the problem was to appear in the DIAL and therefore had to originate the idea that he could locate the gold without the beech.

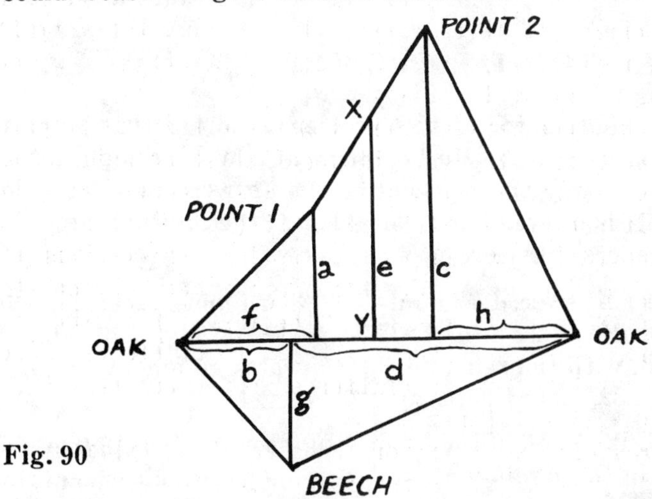

Fig. 90

We have reproduced in Fig. 90 the locations of beech, oaks, points 1 and 2, and point X which marked the spot. By dropping perpendiculars as shown; it is readily evident from the equality of triangles that distance a equals b and c equals d. Distance e which equals half the sum of a plus c must therefore equal half the distance between the oaks. Further, since f and h both equal g, point Y must lie half way between the oaks, so that unmindful of the Missing Beech, Euclid merely stretched a rope between the oaks and walked from its midpoint a distance equal to half the rope length and perpendicular to it.

"However," wrote Maynard W. Teague, of Gulf Research & Development Co., "the story does not end here, for as they prepared to leave who should appear but Silver and his cutthroat pal One Eye Dick! In the cutlass duel which followed, Long John was killed and Silver buried him in the hole where the treasure was found. However, realizing that without Euclid's analytical ingenuity even he could not have found the treasure, Silver spared the boy's life but left him marooned on the island. Euclid's escape is another problem whose solution I hope he was better able to solve than I have been."

22. THE TOASTER PROBLEM

The contributor of this problem reported that when it was presented at the Horlick plant at a Work Simplification Conference only one per cent of the answers received were correct. It is interesting to note that of the solutions received from readers the score was as follows:

Correct Answer, 1.77 min. — 48% of replies received, 1.79 min. — 18%, 2.24 min. — 12%, 1.94 min. and 2.34 min. — 6% each, with the remaining 10% scattered among answers of 1.80, 1.82, 1.90, 1.95, 2.29, 2.37 and 2.44.

Mr. C. W. Yates, Production Engineer for Oldsmobile, did the job in the minimum number of steps, which is ten, and added succinct explanation, as follows: "To fully utilize the toaster, both sides must do an equal share of the work, that is, three sides of the bread slices must be toasted on each side of the toaster. Schedule will be as follows:

Time	Element
.0 - .05 min.	Put in A slice in left.
.05- .10 min.	Put in B slice in right.
.55- .57 min.	Turn slice A.
.60- .65 min.	Remove slice B.
.65- .70 min.	Put in slice C in right side.
1.07-1.12 min.	Remove slice A to plate (done).
1.12-1.17 min.	Put B slice (2nd side) in left side.
1.20-1.22 min.	Turn slice C.
1.67-1.72 min.	Remove slice B (done).
1.72-1.77 min.	Remove slice C (done).

"The reason the second slice put in is removed to make way for the third slice instead of the first put in, is to keep the elements from conflicting and to prevent keeping one side of the toaster idle for .05 minutes longer than the .10 minutes necessary for a change of slices."

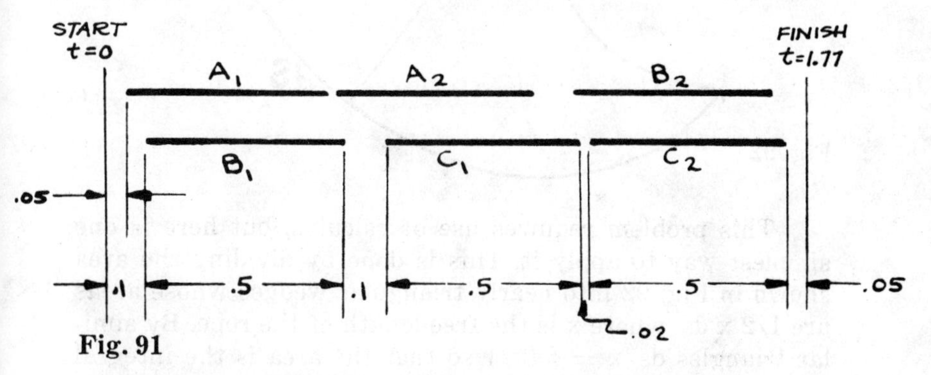

Fig. 91

Mr. Emil J. Rychlik, of Liquid Carbonic Corp., Chicago, contributed a handy diagram on cross-section paper to illustrate his similar solution (see Fig. 91). The many readers who arrived at the figure of 1.79 evidently missed out by turning the second piece instead of the first, which makes the second turn occur in idle time, and others with still higher figures failed to note that if both sides of the toaster were in use as much as possible, the total elapsed time would start from a base of 1.5 min., to which the time for only five removals and/or insertions and one turn would have to be added, as shown in Fig. 91, making the score 1.5 plus .25 plus .02, or 1.77.

23. THE COMPLAINING GOAT

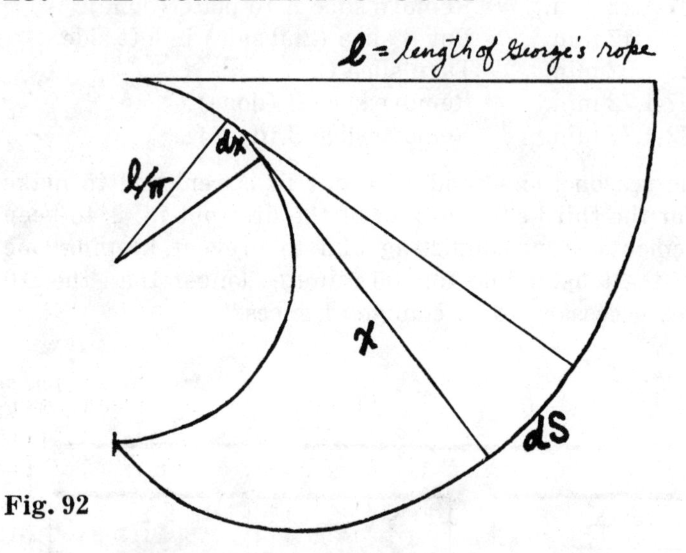

Fig. 92

This problem requires use of calculus, but here is one simplest way to apply it. This is done by dividing the area shown in Fig. 92 into nearly triangular wedges whose areas are $1/2$ x ds, where x is the free length of the rope. By similar triangles $ds/x = \pi dx/l$ so that the area is the integral from 0 to l of $\pi x^2 dx/2l$ or $\pi l^2/6$. George's total grazing area is then $\pi l^2/2 + 2\,\pi l^2/6 = 5\pi l^2/6$ which as compared to Bill's area of $100\pi l^2/121$ is actually $121/120$ as much, or just .833% greater.

24. THE TIMED SIGNALS

It is readily possible to arrive at the answer that any constant speed equal to 30 MPH divided by an odd integer so as to give 30, 10, 6, etc. will avoid "hitting the lights." However, as explained below, this is neither the complete nor the practical solution. To get the figure 30 divided by an odd integer, a helpful procedure used by many DIAL readers is to draw a chart (Fig. 93) between distance d and time t. The horizontal lines are locations of successive signals; the full line indicating red and the dotted line green; the change coming after a time t in minutes equal to (30 x 5280)/60 x space d between the signals (parallel lines) expressed in feet. Any straight line such as drawn on the chart represents a typical constant velocity, 30 divided by an odd integer, will avoid any red signal. However, the problem called for *all* possible constant speeds and if the driver is permitted only the exact figures of 30, 10, 6, etc. this means that he must arrive at all lights including the first at exactly the same interval of time after they had turned green. Such condition is, of course, none too practical nor does it happen to conform to the demands of mathematics unless the assumption is made that there are an infinite number of signals, which presents the intriguing anomaly of an exoteric solution requiring for its correctness an extra-terrestrial situation. Actually, exact speeds of 15, 7-1/2, 3-3/4 etc. are also permissible but equally impractical under certain convenient assumptions (see below). In any event, the whole matter is resolved by drawing a chart in which the abscissas represent the location of successive signals D units apart and the ordinates give the time intervals T. The periods during which each signal is red are shown by the heavy interrupted vertical lines at the signal points. Since, in this case, D/T is 30 mph the D positions are for convenience marked off 12 scale divisions and the T position 4 scale divisions apart. (This gives a ready graphical check of the figures for velocity subsequently derived.) The required answers to the problem are the reciprocals of the slopes of ALL straight lines that can be drawn from the starting point

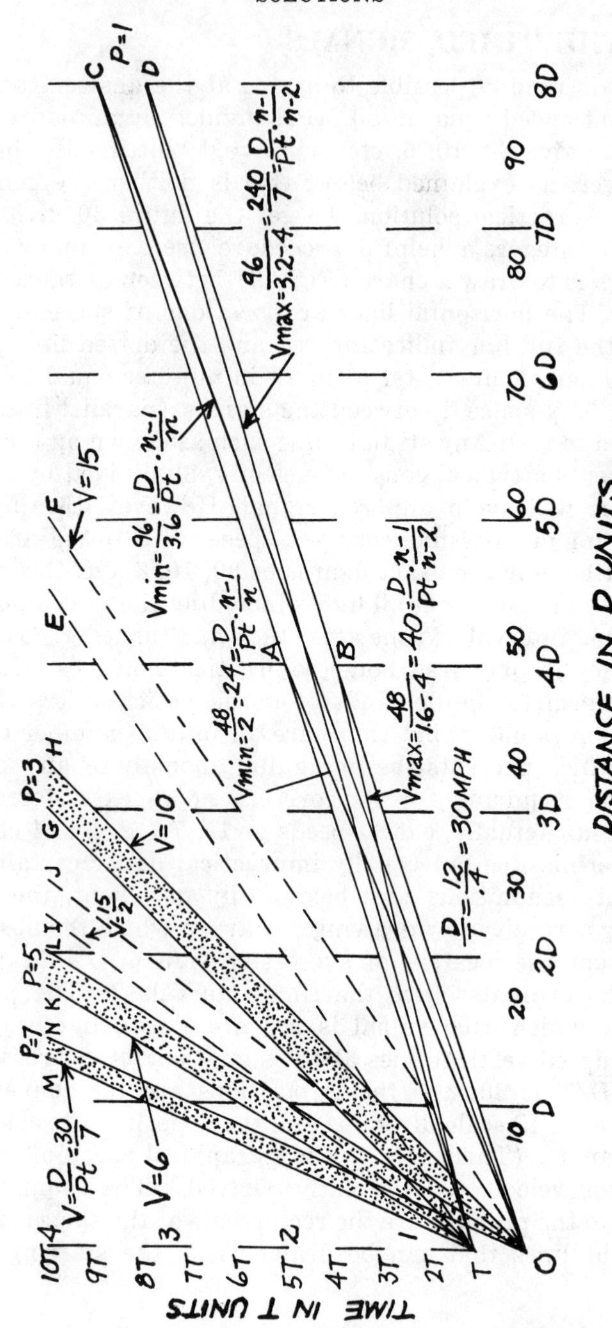
TIME IN T UNITS
DISTANCE IN D UNITS
O
T
2T
3T
4T
5T
6T
7T
8T
9T
10T
D
2D
3D
4D
5D
6D
7D
8D
10
20
30
40
50
60
70
80
90
V=6
V=10
V=15
P=1
P=3
P=5
P=7
A
B
C
D
E
F
G
H
I
J
K
L
M
N

Fig. 93

without intersecting the heavy vertical lines. It is to be noted incidentally that the starting point can be anywhere from 0 to T. The parallel lines TC and OD cover the condition where $p = 1$ in the formula $V = D/pt$ giving a speed of 30 mph. The parallel lines TG and OH are for $p = 3$ and $V = 10$; TK and OL give $V = 6$; TM and ON give $V = 30/7$; etc. Obviously, intermediate speeds are possible within the area TOCD and the other three shaded areas, etc. On the other hand, the parallel dotted lines TE and OF which give the permissible speed $V = 15$ and the corresponding dotted lines TT and OJ for $V = 15/2$, do not admit contiguous speeds since they define relatively "unstable" conditions where the driver must pass one light just after it changes to green and the next just before it turns red.

The influence of the number of signals is readily seen by drawing lines TB and OA for 5 signals and lines TD and OC for 9 signals. These show that in the first case speeds may be had (police permitting) from 40 mph maximum down to 24 mph and for the second case from 34-2/3 (or 30) mph down to 26-2/3 mph, the formulas being $V_{max} = D(n - 1)/pt(n - 2)$ and $V_{min} = D(n = 1)/ptn$ where n is the number of signals and p in this case is 1, a similar set of figures being had for any given number of signals if p is 3, 5, 7 etc. In conclusion, the driver would rarely know just how many signals there were but might base his actions on the likelihood of there being not more than say 12, and if so, he could maintain a constant speed anywhere from 30 mph down to 27.5 mph (which would give him a reasonable margin on his speedometer provided he hit the first signal just after it turned green, with proportionately less leeway if he passed the first signal relatively later. (This set of figures could also be multiplied by 1/3, 1/5 etc., for conditions of blocked road ahead, funeral processions, Sunday traffic, Milwaukee Braves' baseball games, etc.)

25. THE TWO LADDERS

As in most of the problems in this collection, the obvious approach by use of similar triangles proves to be needlessly long and complicated. By similar triangles (see Fig. 94) two

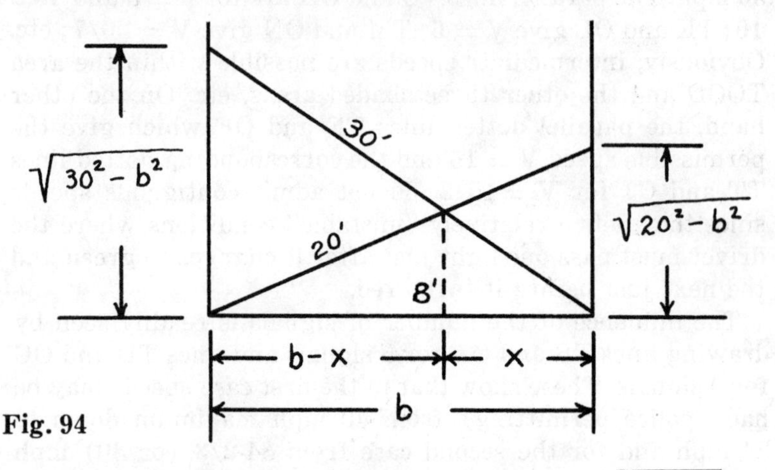

Fig. 94

equations are readily set down $8/(b - x) = (\sqrt{20^2 - b^2})/b$ and $8/x = (\sqrt{30^2 - b^2})/b$. Equating the values of x found from these two equations gives $8/(\sqrt{30^2 - b^2}) = 1 - 8/(\sqrt{20^2 - b^2})$. This biquadratic equation is by no means easy to solve and requires cumbersome successive approximations, plotting by graph, the use of special methods such as Newton's, Horner's, or the recurrent method, all of which were represented in the answers received.

This rather unsatisfactory situation (which, incidentally, led to our selection of the problem in the first place — in the hope that our readers might find a way out) inspired H. G. Taylor of Diamond Chain & Mfg. Company, Indianapolis, an evident devotee of graphic methods, to write us: "The ladder problem is one of my old pets, but I have never seen a short solution for it. However, I have always believed that these problems can be enjoyed by a vastly greater number of enthusiasts if they can be solved without the use of higher mathematics. To that end, quite a number of rather difficult problems can be solved with satisfactory exactness by graphical methods. The ladder problem is a typical example, and the graphical solution can then be checked by the use of "cut and try" algebra." Mr. Taylor proceeded (see Fig. 95) to strike off an arc AB which scales the 20 ft. ladder and then, with a pair of dividers, set to scale the 30 ft. ladder, various ladder-crossing points P are marked off, corresponding to various widths of lane, to form the dotted curve. The abscissa of this curve corresponding to an 8 ft. ordinate gives an approximate graphical solution, which Mr. Taylor

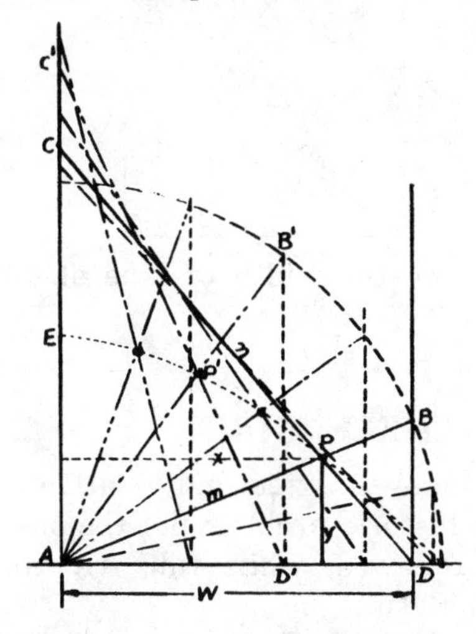

Fig. 95

then applied by one of the trial and error methods to his biquadratic equation, so as to sharpen the result to the required number of decimal places.

Incidentally, the various equations were usually in the form $y^4 - 16y^3 + 500y^2 - 8000y + 32000 = 0$ where y, for convenience, is the height at which the 20 ft. ladder touches its wall or $x/\sqrt{400 - x^2} + 8/\sqrt{900 - x^2} - 1 = 0$ where x is the unknown width. An equation that can be handled in somewhat simpler fashion than either of these was arrived at by Howard D. Grossman, as follows: Let the ladders be divided in the ratio x:y (see Fig. 96) where $x + y = 1$. Then $\sqrt{400x^2 - 64} : \sqrt{900y^2} - 64 = x:y$ which reduces to $64(1/y^2 - 1/x^2) = 500$ or $(x - y)/x^2y^2 = 1/.128$. Now let $x - y = r$ and $xy = s$. Since $(x + y)^2 - 4xy = (x - y)^2$, then $1 - 4s = r^2 = s^4/(.128)^2$ from which $4s = 1 - 61s^4$. Starting with the approximation $S = .2$, the value $S = .2165$, correct to four decimal places, is arrived at fairly readily by loga-

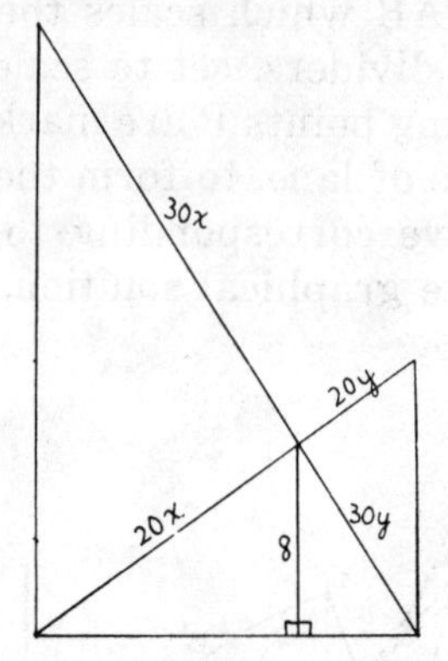

Fig. 96

rithms, from which the width of lane of 16.2 ft is easily derived.

26. THE HATCHECK GIRL

There are many shorter and relatively more erudite solutions to this historic problem but the one most interesting and easily followed, while fully rigorous, requires a fairly lengthy analysis:

Calling the heads A, B, C, D, etc. up to N and the corresponding hats a, b, c, up to n, there are, of course, $n!$ possible distributions of the n hats, out of which there are, say, x_n cases where no hat is on the proper head. Of these "satisfying" cases there is a certain number where hat b is on head A, which number, from the symmetry of the situation, must be matched by an equal sized group in which any other specific hat, — c, d, e, etc. — is on said head A. Therefore, if we could find the number of satisfying cases where hat b is on head A we would merely have to multiply that number by $(n - 1)$ to get the required figure x_n. With hat b ensconced on head A, there are $(n - 1)$ hats left over for $(n - 1)$ heads, with no hat permitted to go on the corresponding head. Let us divide up these cases into two mutually exclusive situations: 1) when hat a is on head B, and 2) when any hat other than a is on head B. The first set of cases merely leaves (n — 2) hats from c to n to go on (n — 2) heads from C to N and this figure, using our previous notation, is x_{n-2}. In the second set of cases we have hats from c to n and heads from C to N with an additional head B that must not be covered by hat a (it is to be remembered that hat b is out of the picture because it is already on head A). Clearly the satisfying cases must be the same as if hat a were here for the nonce called b, the number of such cases being x_{n-1}. This gives the equation $x_n = (n - 1)(x_{n-1} + x_{n-2})$ which can be expressed $x_n - nx_{n-1} = -x_{n-1} + (n - 1)\,x_{n-2}$. It is to be noted that the left side of the equation differs from the right only in sign and in the substitution of n for n — 1. This obviously means that $x_n - nx_{n-1}$ must have the same numerical value for all values of n, although alternating in sign. When n is 2, $x_2 = 1$ and $x_1 = 0$ since there is only one way in which two hats can be put on wrong heads and no way in which one hat can go on a wrong head.

Therefore $x_2 - 2x_1 = 1$ which means that $x_n - nx_{n-1} = \pm 1$ for all values of n, the positive sign applying when n is even, the negative when n is odd. Dividing through by n!, $x_n/n! - x_{n-1}/(n-1)! = \pm 1/n!$ from which $x_n/n!$, which is the required probability equals $x_{n-1}/(n-1)! \pm 1/n!$ This means that for each added hat the previous probability is alternately increased and decreased by 1/n!, an interesting fact which does not evidence itself so readily in other modes of solution. When n is 1, the probability is of course zero; when n is 2 the probability is increased by 1/2! and becomes 1/2; when n is 3, the previous probability is decreased by 1/3! and becomes 1/3; when n is 4 this figure is increased by 1/4! to 3/8; when n is 5, this is decreased by 1/5! to 11/30 and in general therefore $x_n/n! = 1/2! - 1/3 + 1/4!$ ------ 1/n! This is a familiar series that equals 1/e when n is infinite, e being the natural logarithmic base 2.7182818284. . . The series converges rapidly and if n is greater than 10, the probability is actually equal to the true value of 1/e to seven decimal places or .367894, which means that the actual extent of the group checking their hats in the large theater need not be known. Our readers who were good enough (as requested) to check their answers by matching two full packs of cards might just as well have used only a single suit in each. (The totals as reported were 72 satisfying cases out of 210 trials which is reasonably close to the mark.) Incidentally an alluringly short but fallacious approach used by several readers stated that since the chance of any one customer getting the wrong hat is $(1 - 1/n)$, this chance for all customers is that expression raised to the nth power. This would be right if the events were mutually exclusive, which they are not, so that the answer is pretty far off for small values of n, but becomes correct, namely 1/e when n is infinite, when this restriction evidently washes out.

27. GEARS IN THE JUNGLE

Although the problem specified that the gears were twice as wide as they need be from the standpoint of strength, which suggests that they were spur gears (see below), an interesting solution was provided by E. H. Herbert of Oliver, United Filters, Inc., Oakland, Cal., who disdained the extra width of the gears and decided that they were bevels, in which case only nine of them of normal width would be needed, and the construction made much simpler than if they were spurs. His letter follows: "Attached is a sketch (Fig. 97) of the 6 to 1 speed reducer the American engineers built in the Australian jungle with their welding equipment, but you will notice that they had three gears left over and did not have to use the half faces of any of the gears. If the gears had been spurs as the problem probably anticipated, instead of bevels, 12 would have been required but the theory would have been the same. It will be easier to explain the gear train from the slow speed end but that will not prevent the high speed end from being the input. Gear A is held stationary. Gear B rotates around the axis at 1 RPM as its spindle is integral with the slow speed shaft. If the slow speed rotates clockwise, then gear C rotates 2 RPM clockwise. Gear D is attached to C and rotates the same. Gear E rotates on a fixed axis and drives F at 2 RPM counterclockwise. G rotates the same as it is fastened to it. Gear H rotates 2 RPM clockwise around the axis as its spindle is fastened to gears C and D. If G were stationary, then Gear I would rotate 4 RPM clockwise, but as G rotates 2 RPM counterclockwise, it will add 2 RPM clockwise to 'I' making 6 RPM."

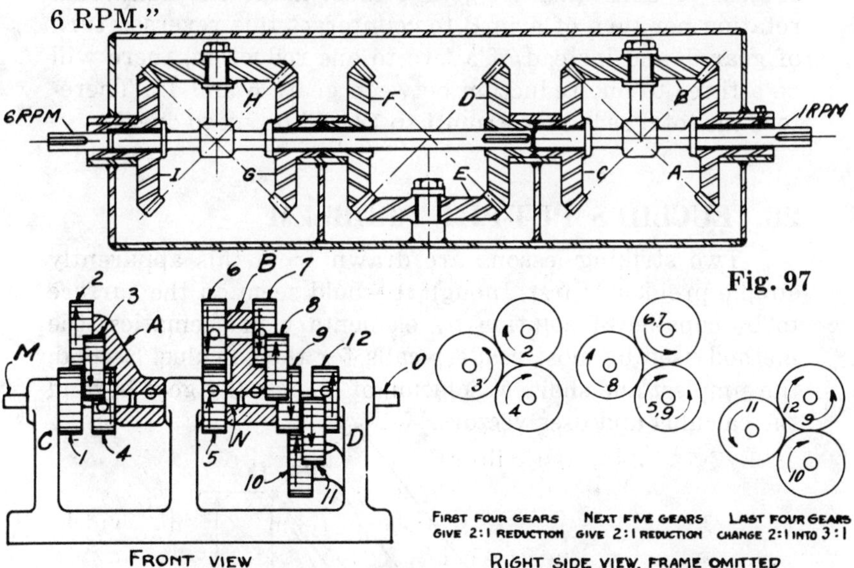

Fig. 97

The more involved solution contemplated by the originator of the problem and which uses all twelve extra width spurs is also shown in Fig. 97: "Gear 1 is fastened rigidly to fixed frame C. Gears 2 and 3 rotate on spindles mounted on the revolving arm A. Gears 6, 7 and 8 rotate on spindles mounted on revolving arm B. Gears 10 and 11 rotate on spindles protruding from the fixed frame D. Gears 4, 5, 6, 7 and 12 are keyed to their shafts. Revolving arm A is keyed to shaft N. Revolving arm B is keyed to shaft O. The rotation of shaft M turns gear 4, and since 1, 2, 3 and 4 are in train, all will try to rotate together. But as 1 is fixed in position, 2 will be made to roll about 1, causing arm A to move along with it and to revolve one turn for every two turns of M, transmitting one turn through shaft N to gear 5. Gears 5, 6, 7, 8 and 9 form a similar mechanism which would give a like two to one reduction if gear 9 were fixed. But through the train 9, 10, 11, 12 it is seen that 9 is not fixed, but that it turns in the opposite direction to that of gear 12, and consequently opposite that of arm B, and likewise opposite that of gear 8. Consequently, gear 5 must make one additional rotation per turn of arm B to counteract this reversed turn of gear 9, and instead of a two to one reduction, there will be a three to one reduction between gears 5 and 12. Therefore the total reduction from 1 to 12 will be six to one."

28. EUCLID'S PUT-PUT PROBLEM

Two striking lessons are drawn from this apparently simple problem. First, though it would seem on the surface to be capable of solution by elementary mathematics, the method used by most readers calls for the calculus. Second, the simplest and shortest solution of all actually goes beyond the calculus and uses vectors.

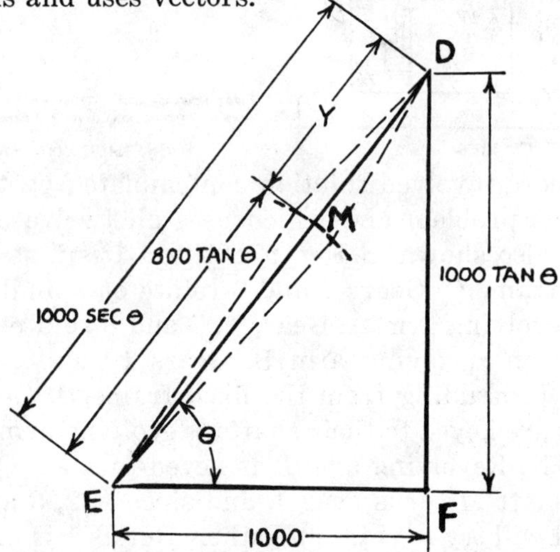

Fig. 98

In the solution by the calculus, it is to be noted first that Euclid's path would have to be towards the point where Pop would be when Euclid got nearest, since a deviation either way from that path would, in the same length of time, have put Euclid farther away (see dotted lines in Fig. 98), a straight line being the shortest distance between two points. The distance y is then readily set up (see triangle EFD in Fig. 98) in terms of the unknown angle as $1000 \sec \theta - 800 \tan \theta$. If Dad's initial and final positions were at F and D, distance FD being $1000 \tan \theta$, Euclid's final position along line ED must have been such that distance EM is 8/10 that amount.

Equating the first derivative to zero in order to find the minimum value of y, we have $y^1 = 1000 \sec \theta \tan \theta - 800 \sec^2\theta = 0$ from which $\theta = \sin^{-1}.8$ and the required distance $y = (1000 \times 10/6) - (800 \times 8/6)$ or 600 yds.

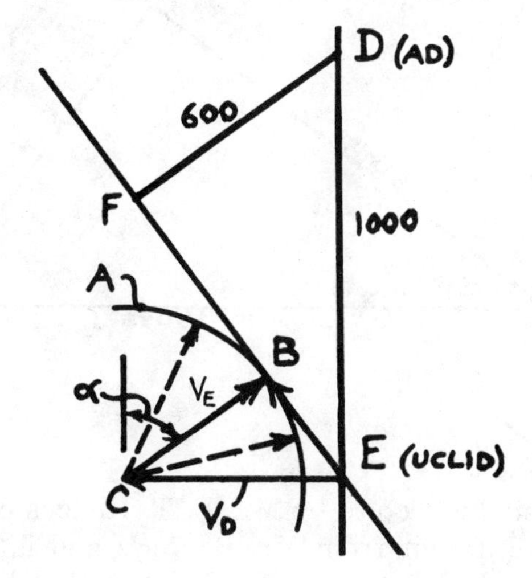

Fig. 99

The above method would have pleased Newton and/or Leibnitz, discoverers of calculus, but it must be stated that a still simpler approach is vector analysis, which became a valuable tool of mathematical physicists in the past century. We quote below the ingenious solution by this method of Meriwether Baxter of Gleason Works, Rochester, N. Y., whose notation, incidentally, is also intriguing: "Considering his fathers' launch stationary, Euclid has a relative velocity V_d shown as the vector EC, and equal to the launch's speed. His own velocity V_e is drawn from C at the unknown angle X. In any event, the terminal point B of V_e lies on circle A centered at C. Euclid's total velocity relative to the launch is the resultant vector EB; the desired solution is given by the limiting case where EB is tangent to circle A, and angle EBC = 90°. (This is because the distance from D to the line of Euclid's velocity EF will then obviously be least). Substituting the given speeds, we obtain angle X = 53°8′ by which little Euclid must depart from his original course; the point of closest aproach is F, reached in about 4-1/2 min., when Junior will have to make himself heard over 600 yds. of water."

Fig. 100

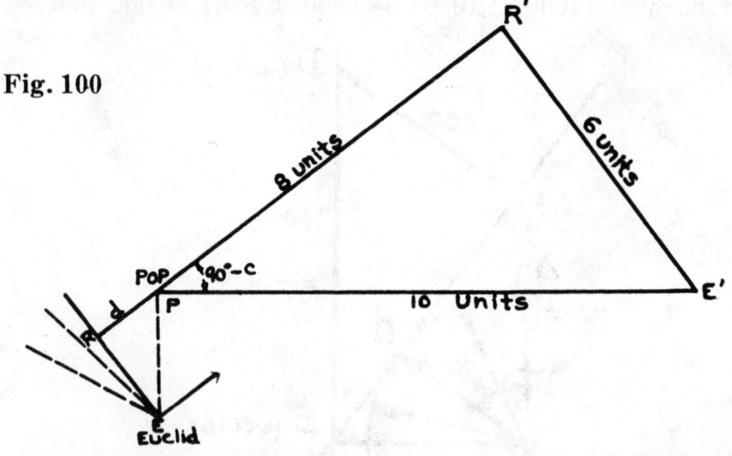

The analysis by vectors by Mr. E. T. Hodges of Norfolk, Va., is a bit different from Mr. Baxter's and is worthy of inclusion. It is to be noted from Fig. 100 that he too took pleasure in choosing the letters in his diagram to match the principles. Wrote Mr. Hodges, "Little Euclid had a brother in the Navy who told him something about maneuvering boards, and Euclid decided he would have to alter his course to the right so that his relative movement line (ER) would be in as northerly a direction as possible. So he drew a line in an easterly direction 10 units long, and swung an arc of 8 units radius from Pop's position. Knowing E′R′ gave him his relative movement line; he saw that he could get the most northerly direction from E by drawing a tangent from E′ to the arc of 8 units radius. This gave him the right triangle PR′E′ where Cos $(90^\circ - C) = \text{Sin } C = .8$. Now he either used a protractor or grabbed a Bowditch and found this to be an angle of 53°8′ (neglecting the seconds). Also being smart in school, he knew a similar pair of triangles when he saw them, and knew that ERP was similar to PRE. Using his trigonometry he found R′E′ was 6 units long and

$$\frac{d}{E'R'} = \frac{EP}{PE'} \text{ or } d = \frac{6 \times 1000}{10} = 600 \text{ yards.}$$

But the story does not end here. Mr. Allan H. Candee, instead of using either vectors or differentiation found it more interesting (though a little more lengthy) to apply geomet-

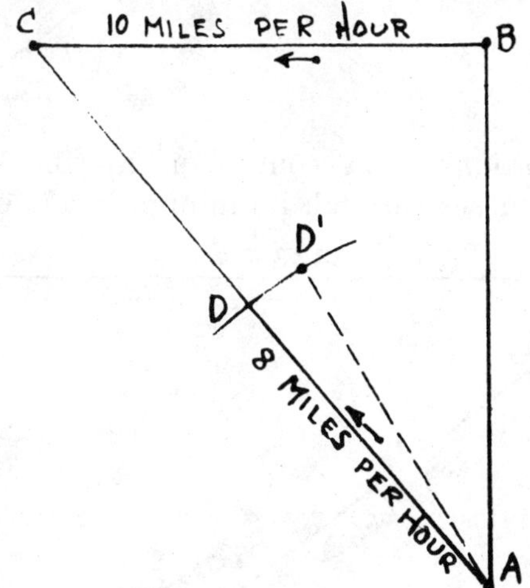

Fig. 101A

rical reasoning as follows: "In Fig. 101A," wrote Mr. Candee, "assume that Pop's boat has traveled from B to C at 10 miles per hour, and that little Euclid steers in the direction from A towards C at 8 miles per hour and reaches D. Then, for the assumed distance BC; DC is the least distance between the two boats, for if little Euclid had steered in any different direction and had thus reached D′, the resulting distance D′C would be greater than DC. (So far this follows the calculus method.)

"Starting with a small distance BC and considering successively increasing distances, the resulting distance DC between the two boats begins by decreasing, reaches a minimum, and after that increases indefinitely. Considering

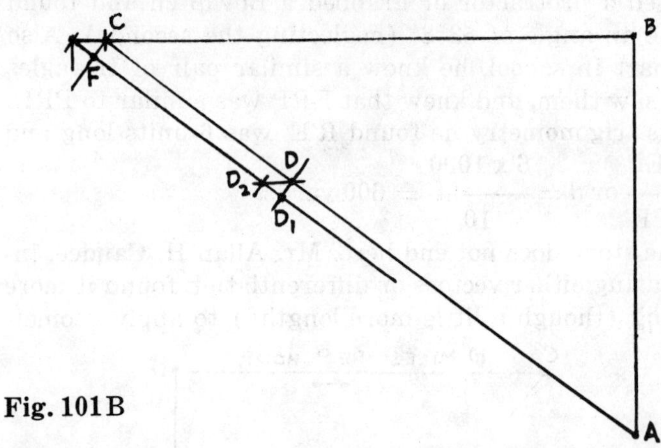

Fig. 101B

these variations to be continuous in time, we know that when DC passes through its minimum value, it is moment-

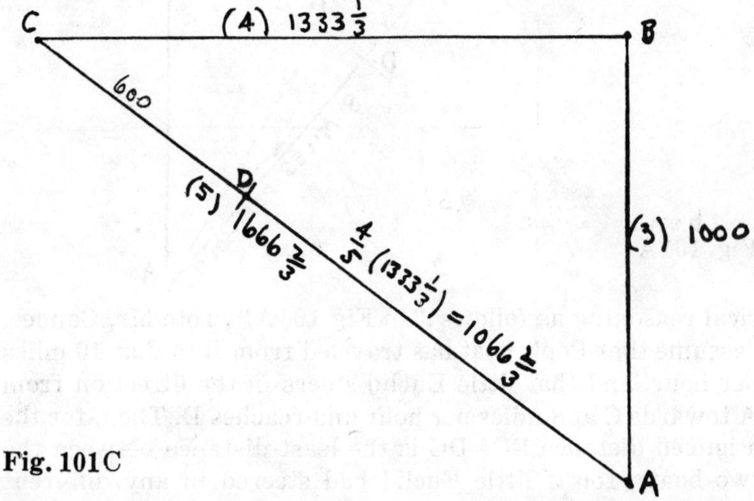

Fig. 101C

arily constant. This consideration enables us to determine the proportions of triangle ABC and the required angle between AB and AC.

"Thus assume that distance BC is the distance traveled by Pop's boat that makes DC the minimum. Then a slight increment CE is accompanied by a corresponding increment FE in the distance AC, and the position of little Euclid's boat changes from D to D_2. Since at the minimum, DC is momentarily constant, D_2E equals DC, and increment D_1D_2 of little Euclid's travel equals increment FE. In the infinitesimal right triangle CEF, CE/FE = 10/8; and in the similar finite triangle CAB, AC/BC = 10/8.

"We now see that the problem has been "rigged" to employ the well-known 3-4-5 right triangle; and it can be solved by mental calculation to obtain the distances indicated in the bottom part of Fig. 101C. The angle by which little Euclid should change his course is angle BAC = $\sin^{-1}(4/5)$ = 53 deg. 8 min. approx., as before."

Another interesting variation in method which also avoided differentiation is shown in the diagram submitted by Mr. Henry Carleton, Special Devices Center, Office of Naval Research, Port Washington, N.Y. Wrote Mr. Carle-

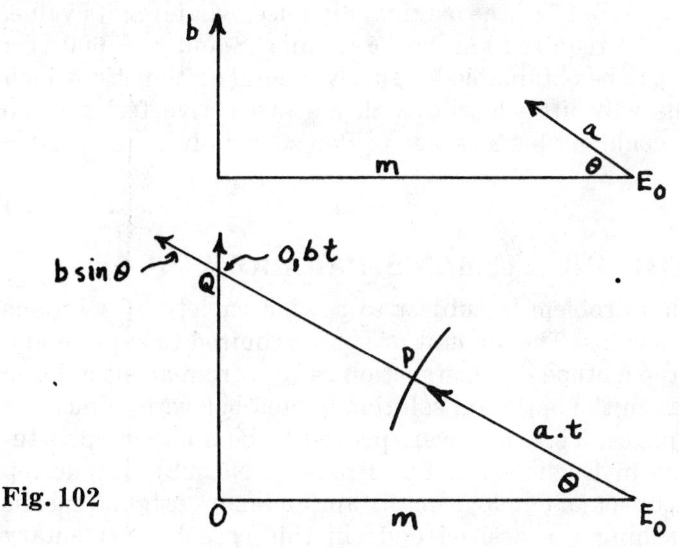

Fig. 102

ton: If we let a = Euclid's (E) speed, b = Father's (F) speed and m = range at t = o, the starting conditions are as shown in Fig. 102.

At time $t = k$, F will have reached the point Q (0, bt). E will be on the circle, center at E_0, radius at. Clearly for this t, a minimum range is achieved by that 0 corresponding to a joining of E_0 and Q.

Then at each time t, the minimum range is the distance from P to Q.

Now consider the rate at which this line is changing. F is lengthening r_{min}, the minimum range line at the rate $b \sin \theta$; E is decreasing its length at the rate a. The desired minimum is obviously obtained when these two rates, of increase and decrease, are equal. From which $b \sin \theta = a$,

$$\sin \theta = \frac{a}{b} = \frac{8}{10} = 53°8'.$$

And to cull the last shred of substance out of this puzzler, it should be noted that the required angle for the change of course is independent of the original distance between pursuer and the pursued. The ratio of velocities being 8 to 10; the third leg of the triangle is of course 6 and the closest approach is 6/10 of the original distance, whatever its value. So that the required answers $\theta = \sin^{-1}.8$ and $y = 600$ yds. turn out to be obtainable by purely mental arithmetic, which explains why little Euclid with his instinctive feel for the elusive could get his message to Pop so rapidly.

29. THE DRAFTSMAN'S PARABOLA

This problem is subject to a wide variety of solutions using calculus. The amount of space required to explain and prove the method of construction ranges from about a dozen lines, as in the optimum solution quoted below, to almost as many pages. This, however, proved to be another opportunity (as in Euclid's Put-Put Problem, No. 28) to sidestep the calculus and employ equally apt and more original means for reaching the desired end; in this case by elementary kinematics familiar to freshman students in engineering. To help appraise adequately the significance of Mr. Candee's achievement along these lines, we will first consider the more conventional methods of solution. It will be remembered that curvature by definition is the rate of rotation of the tangent (or normal) per unit length of arc. The radius of curvature is the reciprocal of this figure, and the center of curvature is the limit of the intersection of adjacent normals. Applying this definition through the calculus to the general curve $y = f(x)$ readily gives for rectangular coordinates the relation for the radius of curvature: $R = (1 + y'^2)^{3/2}/y''$. Where the curve in question is a parabola as in Fig. 103, by Mr. W. W. Johnson, this reduces to the relatively simple equation $R = 2a \sec^3\theta/2$, in which the equation of the parabola is expressed for convenience in polar coordinates as $r = a \sec^2\theta/2$, r being the distance from the focus to any point P.

Now assuming that this was familiar to the harassed draftsman, Mr. Johnson would have him merely mark a point A on the Y-axis the same distance from the vertex O (origin) as P is from the X-axis. Draw PA, which is tangent to the parabola because the origin of a parabola bisects the subtangent at any point, and a line AB through A parallel to the X-axis. Draw PN perpendicular to PA cutting the

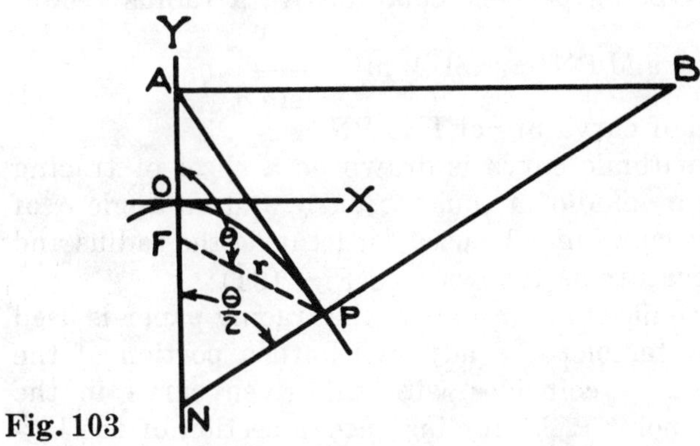

Fig. 103

line at B and the Y axis at N, then BN = R. Note that the focal point F is not necessary in this construction (many readers made the job unduly long by bringing it in) but helps out in the proof, dotted line FP being added in Mr.

Johnson's diagram for the purpose. As above noted, FP $= r = a \sec^2\theta/2$, where θ is the angle AFP. $R = 2a \sec^3\theta/2$ which reduces to $2r \sec \theta/2$. By the properties of the parabola $r = FP = FA = FN = 1/2\ AN$ and $PNO = \theta/2$. Now $NB = AN \sec \theta/2$ (NAB is a right triangle). Hence $NB = 2r \sec \theta/2 = R$.

Another interesting solution employs the logarithmic spiral curve. It is not generally known among draftsmen and engineers that this curve may be employed to advantage for determination of the radius and center of curvature of *any curve* as well as the determination of the evolute.

The unique properties of the logarithmic spiral curve enable one to use it for the rapid determination of the center and radius of curvature of any given curve. It will be recalled that the logarithmic spiral is a curve which cuts the radii vectors from 0 at a constant angle α, whose cotangent is m.

The equation is $r = ae^{m}\theta$

Here a is the value of r when θ is zero. The curve winds around θ as an asymptotic point. If PT and PN are the tangent and normal at any point on the logarithmic curve at P, the line TON being perpendicular to OP, a radius vector, then $ON = rm$ and $PN = r\sqrt{1 + m^2} = \dfrac{r}{\sin \alpha}$.

The radius of curvature at P is PN.

If the logarithmic curve is drawn on a sheet of tracing paper, using a value of a equal to unity and an angle α of about 75°, the curve may be used for locating the radius and center of curvature as follows (see Fig. 104):

The logarithmic curve drawn on the tracing paper is used as an overlay templet and adjusted until a portion of the logarithmic curve coincides with the given curve in the region of the point P. If, for instance, a section of the logarithmic curve coincides with a portion of the given parabola at P, then the line PN, which is tangent to the spiral at N, gives the direction of the true radius of curvature of the point P of the parabola, and the tangent to the logarithmic curve RN, at the right angles to PN, gives the exact location of the center of curvature. Since the constant a equals unity the length of PN may be determined.

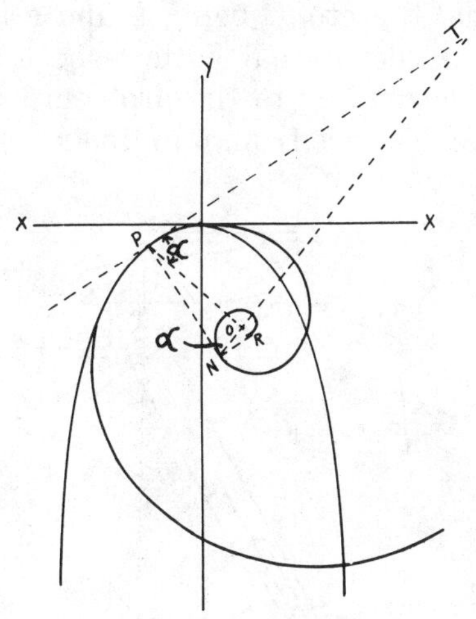

Fig. 104

Note: Assuming that the logarithmic spiral curve has been drawn on a sheet of tracing paper, the construction of the radius and center of curvature requires only two lines, PN and NR.

And now for the solution (see Fig. 105) by kinematics. This approach is prompted by the fact that since curvature (and its reciprocal, radius of curvature) are by definition the rate of relative motion of geometrical entities, these entities can be replaced by links and the desired rate found by the familiar method of instant centers. Five construction lines are needed (the five dotted lines shown in the Fig.) which are two more than in the above version, but all knowledge of calculus is eschewed — requiring merely a familiarity with elementary kinematics plus two well-known properties of the parabola — first, that the origin bisects the subtangent, as mentioned in the first solution above, and second, that the subnormal is a constant length. PQ is drawn parallel to YY. t is laid off so that Ot = y and tangent tP drawn. Normal Pn is drawn intersecting YY at n and nQ is drawn perpendicular to Pn. QC is drawn parallel to xx and intersects the normal at C. PC is the required value.

The proof is based on the conception of three links shown in the heavy lines, two of which are the parabola and the swinging normal thereto, C being found as the instant center of motion of the normal with respect to the parabola. The third link is the key to the problem and makes use of the fact that the distance from n to line xx remains constant.

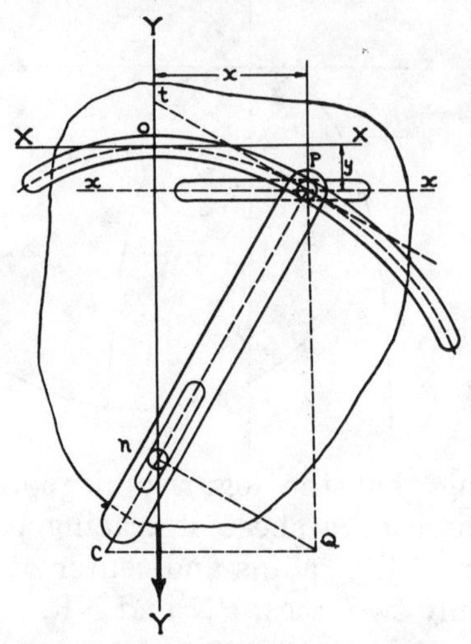

Fig. 105

Conceive, first, a fixed plate with a parabolic slot, in which slides a loose pin P; second, a moving plate with a fixed pin N screwed into it (here is where the constant subnormal comes in) as well as a horizontal slot xx into which pin P projects; and third, a rod having a hole at one end to fit over pin P and a slot at the other to slide over pin n. The second plate is moved directly downward, causing pin P to move outward in xx along the parabolic path and thus determine the path of rod nP which is the normal. The instant center of motion of the rod with respect to the moving plate is found at point Q. This is because point P moves horizontally with respect to the plate and must therefore have its center in PQ, whereas point n which moves along PC must have its center in nQ which is perpendicular to PC. Q is at the intersection of the two lines. The instant center of motion of the moving plate, which has a straight line downward motion with respect to the fixed parabola, is on a horizontal line at infinity, so that the required instant center of motion of the normal with respect to the parabola, which is obviously on the line Pn, since Pn is perpendicular to Pt, must also be on a horizontal line drawn through Q, or at point C, because the three instant centers must be on one straight line.

Mr. A. H. Candee, author of this most interesting problem, which is somewhat more advanced than most in this book, added this significant comment: "In my experience geometry is especially useful to designers. I am not well acquainted with the history of mathematics, but I feel sure that something about curvature must have been known before the introduction of analytic geometry and the calculus. Consider all the curves named after old Greeks." Mr. Candee, incidentally, provided an alternative sketch (Fig. 106)

Fig. 106

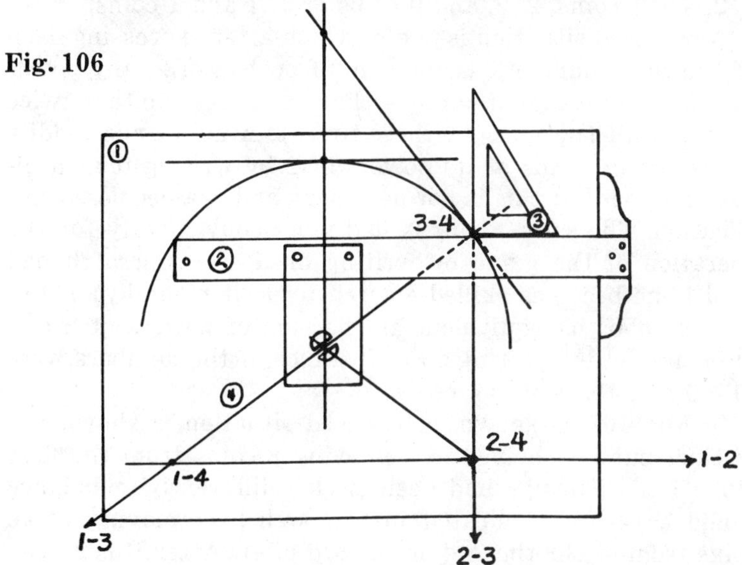

of special appeal to the draftsman, since the parabola is drawn on the drawing board (1), a T-square (2) having an attached piece extending below and carrying a slotted pivot on the parabola axis, and at a distance from the top of the T-square equal to the constant subnormal, serves as the vertically moving upper plate, and (3) a triangle (which is here analyzed for instant centers as a separate link) serves as pin B, being slidable on the T-square and pinned to a rod (4) that passes through the slot. All instant centers are readily determined as shown and the length from (1.4) to (3.4) is the required radius of curvature, as before.

30. VARIATION OF THE GAME OF NIM

Briefly expressed, the winning procedure for A may be stated: Confront your opponent B — if he gives you the chance — with a balanced situation, which will force him to unbalance it, and keep doing this until an end condition is reached where you obviously must win; conditions which include the case where it's B's move and there are either 3, 2, and 1 coins, 2, 2, and 0 coins or 1, 1 and 1 coins.

A balanced situation is one in which after expressing each of the three numbers as the sum of powers of 2, any power that is present is paired off — that is, it occurs either twice or not at all (not once or three times). A number is readily expressed in terms of its powers of 2 by writing it as a binary number but this is not necessary and is a needless complication. The same result is had more conveniently for the operation of the game by writing on a secret scratch pad (until one becomes skilled enough to do it mentally) below each number its equivalent as the sum of powers of 2 and then circling the pairings. For instance, if the numbers were 31, 19 and 15, write as below:

In Fig. 107 as shown, a balanced situation is shown not to exist but is achieved by removing 3 coins from the first pile (if the pairings had been circled differently, a balance would have been indicated just as well by removing three coins from either the 2nd or the 3rd pile). After B is forced to unbalance the situation, A again puts it in balance and so on until one of the three end situations is reached.

A few simple auxiliary rules are more or less obvious and help confound the loser by rapidity of action: 1) If the highest power of 2 appears in the largest pile only, the removal must obviously be from that pile. 2) If at any time poor B leaves two equal piles, remove the third pile completely and then duplicate B's moves unless he leaves one in one pile, in which case you, of course, remove the other pile completely, or unless he takes an entire pile, in which case you take all but one of the remaining pile (in other words, temper the rule of the balanced situation by common sense).

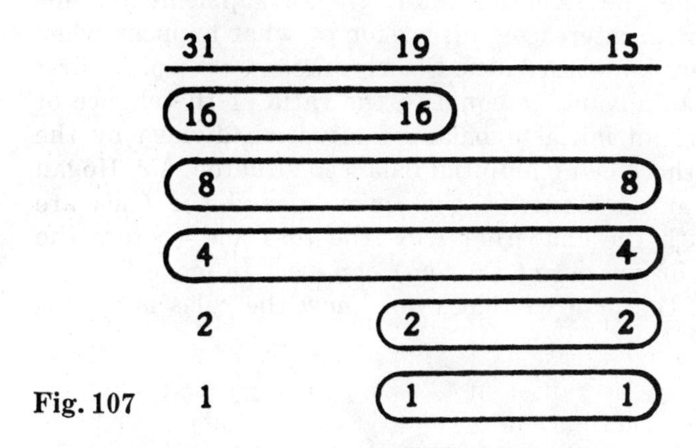

Fig. 107

Incidentally, for those who want to become adept at this pastime, perhaps the simplest way to dispense with the scratch pad and learn how to do the job mentally is to check each number rapidly for each power, starting with the highest. This usually tells with relative ease which of the three piles must be drawn from. Obviously the number remaining in that pile after the draw must equal the sum of the unbalanced powers in the other two, which narrows the consideration to two piles and gives a quick determination of how many coins to take. For instance, if the three piles are 24, 13 and 11, there is only one 16 present, so the draw must be from the first pile. Eight and one are each included in the next two piles, so the unbalanced powers total 6, which means that 18 must be taken from the first pile.

Mr. Joseph T. Hogan of Southern Regional Research Laboratories, reminds us that this game was once a favorite among the habitues of the Creole coffe shops in old New Orleans. Prof. Bouton of Harvard College proposed the name

Nim in 1901. It has often times been called Fan-tan, but it is not the Chinese game of that name. Certain forms of the game seem to be played at a number of American universities and country fairs. One form of the game specifies that the player taking up the last coin or counter from the table wins, the only difference being in the obvious change of handling of the end situations. Mr. Hogan supplemented his solution by an interesting discussion of what happens when both players know the rules. Clearly, the one who plays first will have an advantage equal to the ratio of the chance of there being an initial unbalanced situation divided by the chance of there being an intial balanced situaton. Mr. Hogan went so far as to derive the formula, and the odds are heavily with the chap that wins the toss and (before the coins are thrown out at random) can elect to move first. In any event, the man who does not know the rules is always out of luck.

31. EQUIDISTANT MEETING POINTS

It is surprising to find that many readers who solved this problem when it appeared in the DIAL (possibly you slipped similarly) jumped to the convenient conclusion that the house was equidistant from the three sides of the triangle. To reach this conclusion one reader wrote with great plausibility: "If the points of intersection of the driveways with the roads are equidistant from each other, then they must be the vertices of an equilateral triangle which is inscribed in the triangular piece of land. The center of a circle which is inscribed in the piece of land and which circumscribes the equilateral triangle must be the location of the house for then the driveways will be perpendicular to the roads." This approach gives an answer of 274.8 rods for the radius of the inscribed circle whose center is of course located at the intersection of the three bisectors of the angles of the triangle, and if the three chords of the circle each subtended arcs of 120° the sides of the equilateral triangle would then be 274.8 times $\sqrt{3}$ or 476.3 rods, which happens to be fairly close to the correct answer of 481.3 rods. The fly in the ointment, however, is that the circle inscribed in this particular triangle is not the same as the circle passed through the three meeting points. This is because the three chords of the circle inscribed in a scalene triangle do not subtend angles of 120° but do subtend angles respectively equal to 180° minus the opposite angle of the triangle. We therefore cannot make any use of the inscribed circle in the solution but must resort to other methods.

It is not hard to find that there is enough information to set up 3 simultaneous trigonometric equations, and several readers (as well as the author of the problem) went about the thing in that arduous way by trigonometry or analytical geometry, consuming several sheets of equations and calculations in the doing. Two readers, Mr. Ed. Prevost of Beauharnois, Quebec and Mr. Howard D. Grossman of New York,

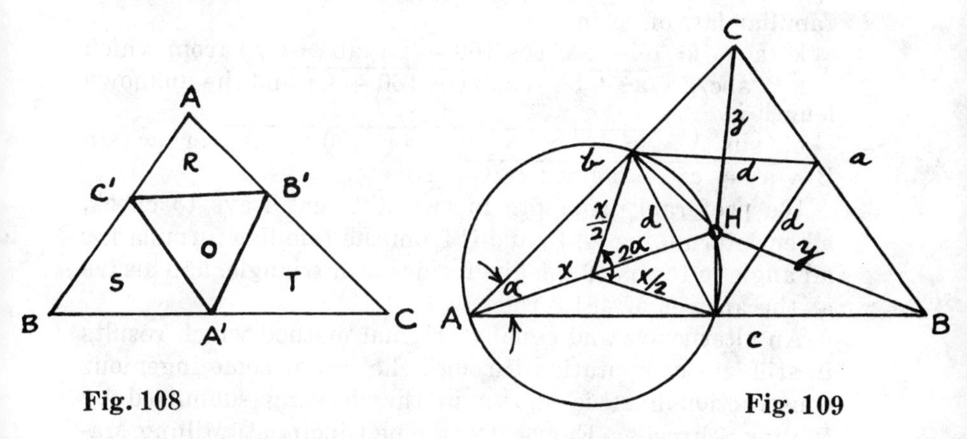

Fig. 108 Fig. 109

hit upon the same ingenious device which shortens the job and gives it a fascinating touch. Triangle ABC has sides a, b, and c. Let the equilateral triangle be A′B′C′ with OA′ perpendicular to BC, OB′ to AC and OC′ to AB. Let R. S and T be the centers of the circles circumscribing the small triangles in which R, S and T lie. These three points turn out to be very useful and are the key to the method. To begin with, since an inscribed right angle must subtend a semicircumference, lines drawn perpendicular respectively to AB at C′ and to AC at B′ must intersect on the circumference of the circle whose center is at R, which means that point O which, by the conditions of the problem is the meeting point of these perpendiculars, is on the extension of radius AR, RO being equal to AR. (The same, of course, applies to the other two circles).

It follows from similar triangles that RS is parallel to and equal to one-half of AB, and is perpendicular to OC′. Since the radius of the circle circumscribed about a triangle equals any side divided by twice the sine of the opposite angle, C′R must equal C′B′/2 sin A. For convenience, let C′B′, the unknown length, equal 2k sin A/a where k is to be determined. Then C′R equals k/a and similarly C′S = k/b. Angle C′RB′ is twice angle A and angle C′CA′ is twice angle B, therefore angle RC′S equals 240° − angle A − angle B or angle C + 60°. We can now determine k from triangle RC′S from the familiar law of cosines.

$k^2/a^2 + k^2/b^2 - 2k^2 \cos(60 + C)/ab = c^2/4$ from which

$k = abc/2\sqrt{a^2 + b^2 - 2ab \cos(60 + C)}$ and the unknown length is

$bc \sin A/\sqrt{b^2 + c^2 - 2bc \cos(60 + A)}$ or $ac \sin B/\sqrt{a^2 + c^2 - 2ac \cos(60 + B)}$ etc.

We preferably compute in two different ways to check, after determining A, B, and C from the familiar formula for an angle in terms of the given sides of a triangle, and arrive at the answer of 481.3 rods.

An alternative and equally original method which results in still less computation through the use of some ingenious construction lines is shown in the drawing submitted by Walter Schroeder, Engrg. Dept. The Cincinnati Milling Machine Co., Cincinnati (Fig. 109). Here, line AH is joined and a circle is drawn about AH as diameter, which circle will of course pass through two of the "meeting points." Radii are joined as indicated; the angle at the center is obviously equal to 2α and it may be shown that $d = 2 \cdot x/2 \cdot \sin \alpha = x \sin \alpha$. Since the area F of triangle ABC $= 1/2\ c \cdot b \cdot \sin \alpha$, $\sin \alpha = 2F/bc$ and $x = dbc/2F$. Similarly, $y = dac/2F$ and $z = dab/2F$. From these relationships it is found that $d=2F/\sqrt{2\sqrt{3}F+\frac{a^2+b^2+c^2}{2}}$. F is calculated from Hero's formula and the value of d is found to be 481.3 rods, as before.

32. CUTS IN A LENGTH OF WIRE

This seemingly simple problem, when it appeared in the DIAL, attracted an unprecedented variety of solutions by arithmetic, calculus, geometry, analytical geometry and algebra. The problem consists in finding the chance that none of the three sections be greater than half the total length, since no side of a triangle can be larger than the sum of the other two. Any approach by arithmetic is by consideration of averages and therefore not a "rigid" solution. In this approach it is reasoned that the second cut must be in the larger section left by the first cut, which section, since it could be anything between l/2 and l, could be said to have an average length of 3l/4, whereby the chance of the second cut being therein is 3/4. To avoid having the second cut leave a length greater than l/2 at either end of the "average" 3l/4 piece requires that it be in the middle third, making the total chance 3/4 times 1/3 or 1/4. Although this reasoning gives the correct answer, its soundness might be queried. It will be remembered that reasoning by averages led to a fallacious result in "Start of The Snow" (No. 12).

The basically correct procedure in a problem of probability like this is to consider all favorable happenings and divide by the total possible happenings. This can readily be accomplished by calculus, the method used by many readers. Although complicated, a solution using this method is included for its interest. Let the length of fence be represented by the line a. Let x represent any arbitrary length of the fence from the left end, x being less than 1/2a and dx being its increment. The probability that the first cut will land on the portion of the line indicated by dx is dx/a. The second cut must be made so that it will land on the heavy part of the line to the right of center, that is, it must be less than 1/2a from the right end and less than 1/2a from x. Otherwise no triangle could be made, since any side of a triangle must be less than half the total of the three sides. But the heavy part of the line is also equal to x. (Since the heavy line $= a - 1/2a - (a - 1/2a - x) = x$.) Therefore the probability of the second cut landing on the heavy line is x/a. The probability that both cuts will occur as specified is xdx/a^2. Summing these from 0 to 1/2a we obtain

$$\frac{1}{a^2} \cdot \int_0^{\frac{1}{2}a} x\,dx = \frac{1}{a^2} \cdot \frac{1}{2} x^2 \Big|_0^{\frac{1}{2}a} = \frac{1}{a^2} \cdot \frac{a^2}{8} = \frac{1}{8}$$

Since the probabilities from 1/2a to a would amount to the same (working from the right end), the total probability would be 1/4.

This particular problem lends itself more simply to geometric solution than to calculus and was contributed by Phelps Meaker, G.E. Co., Cleveland. In a square with side P, he drew diagonal lines as shown (Fig. 110) to the midpoints

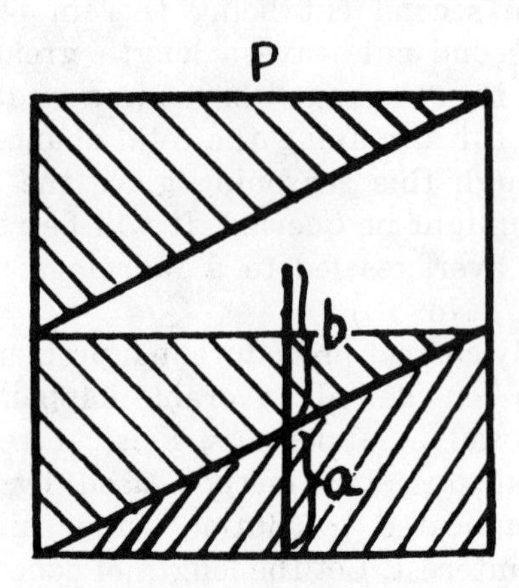

Fig. 110

of opposite sides and then let a random point represent the length a of the first cut on the lower diagonal at a distance a above the base. Clearly, the second cut at a distance b above the first point would have to come within the unshaded area to avoid a section greater than 1/2, so that the favorable situations constitute one-fourth of the total. Another geometric analysis of this kind which followed the historic solution by Poincare, was offered by Joseph T. Hogan, Chemical Engineer, New Orleans, and Mr. W. Weston Meyer, Bundy Tubing Company, Birmingham, Mich.

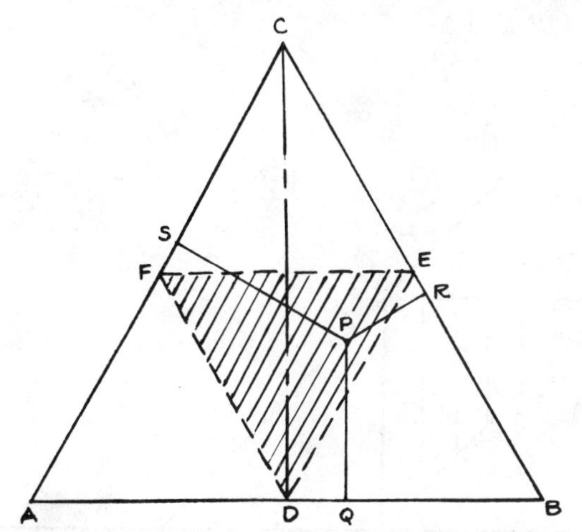

Fig. 111

They used an equilateral triangle with altitude 1 equal to the length of the fencing. By familiar geometry it is known that the sum of the lengths of perpendiculars dropped from any point P to the three sides add up to the altitude, and in order for none of these perpendiculars to be greater than 1/2, P must be within the shaded triangle (Fig. 111), which is one-fourth the total.

Solutions by analytical geometry included one from Thomas L. Rourke, Project Engineer, Ecusta Paper Co., Orange, Conn. who drew the lines $x + y = 1$ and $x + y = 1/2$ (Fig. 112) and reasoned that if the three cuts are x, y and $1 - x - y$, the point P defining cuts x and y, then the third cut is the horizontal or vertical distance therefrom to the line $x + y = 1$. Thus, the favorable location of P defines an area equal to one-fourth of the total.

However, the optimum solution was furnished by Samuel M. Sherman, Philadelphia, whose solution was so general as to apply not only to the chance that two cuts would provide lengths to form a triangle, but also to the chance that three random cuts would form a quadrilateral (where, as before, no length could be greater than 1/2 the total length). It applies also to the chance that n cuts in a piece of unit length would leave none of the (n + 1) pieces with a length greater than x, where x could be assigned any specific value between 1/2 and 1.

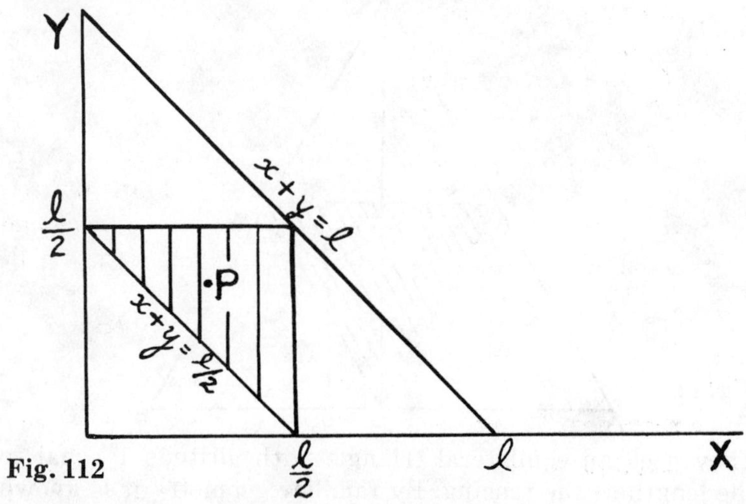

Fig. 112

This general problem was best solved by Robert A. Harrington, B. F. Goodrich Research Center, Brecksville Ohio, who wrote: The problem of n random cuts in a wire of unit length is the same as the problem of (n + 1) random cuts in a ring of unit circumference, except that the latter can be treated more easily because of symmetry. Consider the segment clockwise from the m'th cut in the ring. The probability that none of the other n cuts is as near as distance x from this cut, in the clockwise direction, is $(1 - x)^n$. This is the probability that this particular segment is longer than x. The same holds true for the other n segments. Now if x is 1/2 or more, only one of the segments can be greater than x; that is, the probabilities for the segments are mutually exclusive and the probability that any one segment is greater than x is the sum of the equal individual probabilities, or $(n + 1)(1 - x)^n$. The required probability that none of the segments is greater than x is $1 - (n + 1)(1 - x)^n$. When $x = 1/2$ this becomes the probability that the segment lengths can form a closed polygon. For three cuts in the straight wire $n = 3$, and the probability that the resulting pieces make a quadrilateral is $1 - (3 + 1)(1 - 1/2)^3$ or 1/2. It is interesting that when $x = 1/2$, one gets the correct probability, zero for both $n = 0$ and $n = 1$. Mr. Harrington went on to analyze in interesting fashion the case when x is less than 1/2, the probabilities then not being mutually exclusive, but this takes us a bit beyond the confines of this volume.

33. THE TANGENCY PROBLEM

The conventional solution makes use of a helper circle of any convenient radius as shown in Fig. 113. The angle between the two intersecting lines is bisected by line OG.

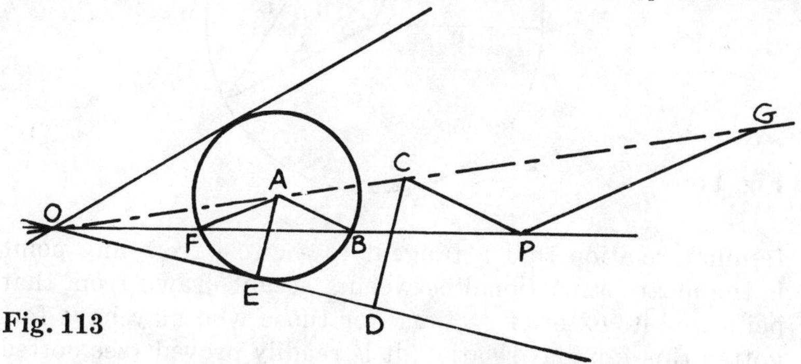

Fig. 113

Any convenient point A is chosen on OG and a helper circle is drawn in the usual way, tangent to the two lines (the distances from A to the two lines being obviously equal). OP is drawn through the given point P intersecting the helper circle at F and B. CP and PQ are drawn parallel respectively to AB and AF. C and G are the centers of the two required circles, as is readily proved by dropping perpendicular CD, from the two sets of similar triangles OAE and OCD; and OAB and OCP. Ratio OC/OA equals PC/AB and also equals

CD/AE, and since AB and AE are equal, PC and CD are likewise equal and the circle with center C will go through P.

Mr. Groen of Onsrud Machine Works, Chicago, showed that the auxiliary circle could be dispensed with in an alternative, though somewhat longer, solution which uses the

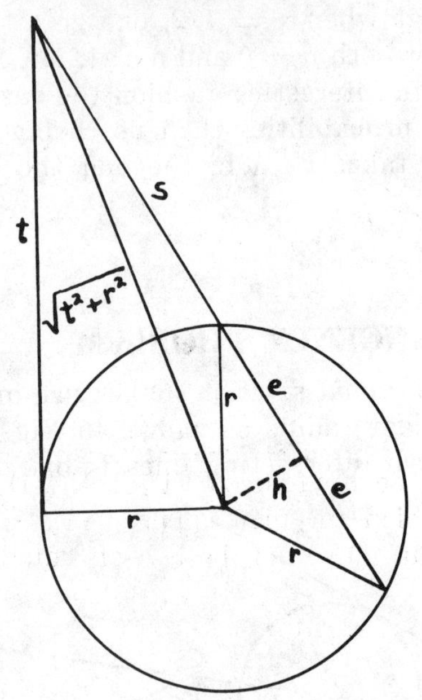

Fig. 114

familiar relation that a tangent to a circle from any point is the mean proportional between a secant drawn from that point and its exterior section. For those who may have forgotten this familiar relation, it is readily proved (see dotted lines in Fig. 114) by joining the external point with the center of the circle, drawing the three radii as indicated and dropping the perpendicular from the center of the circle to the secant. Then, from right triangles, $t^2 + r^2 = (s + l)^2 + h^2$ and since $h^2 = r^2 - l^2$, $t^2 + r^2 = (s + l)^2 + r^2 - l^2$, from which $t^2 = s^2 + 2sl = s(s + 2l)$, which was the relation to be proved. Applying this relation to our problem, line CD is drawn through point A perpendicular to the bisector of the angle (see Fig. 115), a semi-circle is constructed on CD, and the perpendicular AG laid off on line DF equal to DE. Perpendicular EO locates the center of the required circle (the other one is found by laying off DE on the other side of point D). For proof — in order for the circle to pass

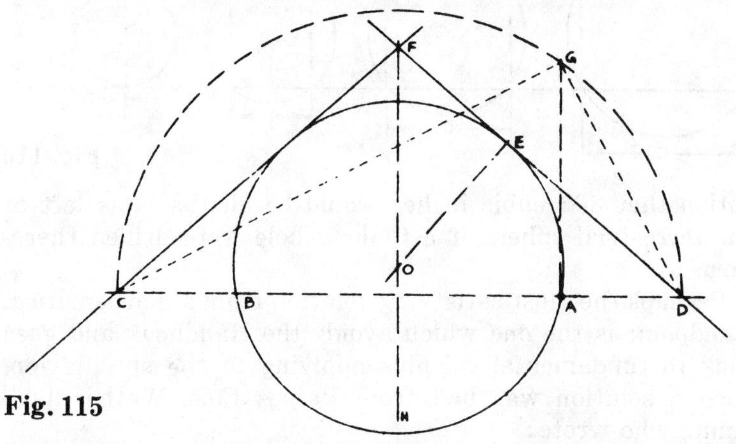

Fig. 115

through A and be tangent to FD at E, $(DE)^2$ must equal (DB) (DA) which in turn equals (AC) (DA) or $(AG)^2$, which means that DF must be laid off equal to AG, as was done.

34. HOLE IN A SPHERE

Most of the solutions received applied the familiar handbook formulas for the volume of the cylinder removed by the drill and the volume of the two removed spherical caps or segments. On subtracting the sum of these from the volume of the sphere the terms involving both the radius of the drill r and the radius of the sphere R cancelled out leaving a remainder $(4\pi/3)\,(h/2)^3$ which is the volume of a sphere with diameter equal to the length of the hole. This of course, would have to be the case if the volume left were independent (as it evidently was) of the radii of the sphere and hole since in that event it would have to be the volume left with a 6″ sphere and an infinitesimal hole. Frank L. Bracy, Jr.,

Plax Corp., Hartford, illustrated this relation as per sketch in Fig. 116, and several readers made the interesting obser-

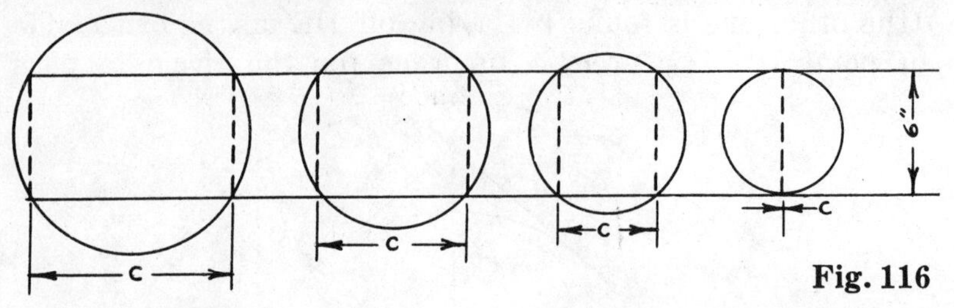

Fig. 116

vation that 36π cubic inches would be all that was left of our terrestrial sphere if a 6″ long hole were drilled therethru.

Perhaps the most satisfying solution from a mathematical standpoint is the one which avoids the Handbook and goes back to fundamental calculus applying to the specific job. Such a solution was had from Phipps Cole, Wethersfield, Conn., who wrote:

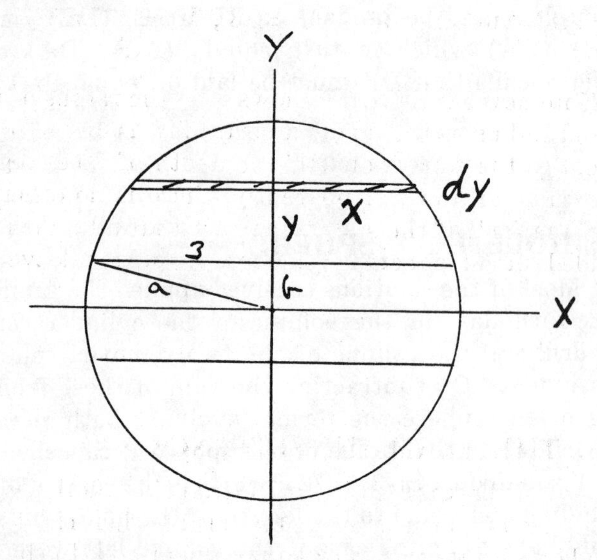

Fig. 117

Draw a cross-section of a sphere (Fig. 117) of radius a with hole 6 inches long of radius b. Note that $a^2 = b^2 + 3^2$. The volume of the elemental cylinder whose cross-section is indicated is: $v = 2\pi y\ (2x)\ dy$ and the volume left in the sphere is the integral of this from b to a:

$$\text{So } V = 2\pi \int_b^a 2xy\, dy \text{ but } x = (a^2 - y^2)^{1/2}$$

$$V = 2\pi \int_b^a (a^2 - y^2)^{1/2}\, 2y\, dy =$$

$$-\frac{4}{3}\pi (a^2 - y^2)^{3/2} \Big|_b^a$$

$$= \frac{4}{3}\pi (a^2 - b^2)^{3/2}.$$

(This is the familiar volume of a sphere $= \frac{4}{3}\pi a^3$ when $b = 0$.) But since $a^2 - b^2 = 3^3$, $V = \frac{4}{3}\pi 3^3 = 36\ \pi$ cubic inches.

Note how adroitly this shows that the radius of the sphere a and radius of the hole b appear only as $(a^2 - b^2)$ which is equal to the square of half the depth of hole, regardless of the value of a or b. Apparently, there is no other "physical explanation" of the surprising fact that all 6 inch long holes drilled in all spheres 6 inches or bigger leave the same residue.

35. THE FARM INHERITANCE

A little knowledge of Diophantine methods and equations saves much time in finding another right-angled triangle with integral sides and of equal area to the triangular plot bequeathed to little Euclid, which had sides of 210, 280, and 350 rods. Obviously the problem resolves itself into solving two simultaneous equations with three integral unknowns, $x^2 + y^2 = z^2$ and $xy = (210)(280)$, which was the particular specialty of Diophantus, the great algebraist who lived about the middle of the third century.

A first step in such procedure might be to recognize that in the simplest group of right triangles (Diophantus set up parameters which established all such triangles) the smaller leg is any odd number, such as 7, the larger leg is the square of that number, less one, divided by two, such as $(7^2 - 1)/2 = 24$, the hypotenuse then being the larger leg plus unity, or 25. (Another group of right triangles is of course had by multiplying each side of a triangle in the above group by the same factor, such as $3 \times 70 = 210$, $4 \times 70 = 280$, $5 \times 70 = 350$, which are the sides in Euclid's inheritance.)

Since the two legs of the unknown triangle must here multiply to give 210×280 so to have the required equal area; the obvious first step is to test two factors of that product, to see if the square of one is about double the other. The factors 49 and 1200 fill the bill exactly since $(49^2 - 1)/2$ equals 1200; the hypotenuse then being 1201, and it is not difficult then to show that no other pair of factors will serve.

However, there is more to be extracted from this problem. A question worthy of consideration is: how did the farmer know that there actually would be another equivalent triangle when he picked the first one? In other words, is there a way to preselect right triangles of equal area and integral sides? On this point, Mr. Edmunds, contributor of the problem, reminds us that Pierre Fermat in 1640 discovered that if a, b and c are integral sides of a right triangle, in which case $2ac(b^2 - a^2)$, $2bc(b^2 - a^2)$ and $2c^2(b^2 - a^2)$ are obviously integral sides of a second right triangle, then $4abc^2$, $c^4 - 4a^2b^2$ and $c^4 + 4a^2b^2$ are integral sides of a third right triangle with area equal to the second. The farmer merely needed to be familiar with Fermat to draw up his will and naturally took the simplest case 3, 4, and 5, and it is assumed that little Euclid, being equally familiar with Fermat, quickly applied the above relationship to find the answer.

If Dad had really wanted to make the thing difficult he could have picked a triangle with sides of 1320, 518 and 1418 and asked Euclid to find three other right triangles of equal area: 280, 2442 and 2458; 2960, 231 and 2969; and 6160, 111, and 6161.

36. THE CHINESE GENERAL

The location of this problem was probably chosen because the general method of solution is based on the familiar Chinese remainder theorem involving the use of so-called congruences. However, in this particular case ,the figures were such that those unfamiliar with that branch of number theory could short-cut the job, as was done by readers of the DIAL in many ingenious ways.

Of these, perhaps the shortest was contributed by W. W. Bigelow, of A. O. Smith Corp., who reasoned from the remainders before and after; that N, the number lost, must equal $4a + 2$ and also $5b$, $7c + 1$, $11d + 10$ and $17e + 16$. (Note that by proper selection of parameters, minus signs are avoided.) The first two of these disclose that N is even and a multiple of 5, specifically an odd multiple of 10. By inspection, $N = 50$ also satisfies the third and fifth relations but not the fourth. Since 4, 5, 7 and 17 have no common factor, N must then equal $50 + 4.5.7.17\ f$, which leaves only the fourth relation to check for successive trial values of f. Taking $f = 1$, $N = 2430$ which also satisfies the fourth relation. So that the general lost 2430 soldiers and had 2851 left.

Actually, it is not necessary in the general method to consider the situation prior to the battle at all, except insofar as the original total gives a key to the most likely answer, out of the infinite number of possible solutions. The general solution that may be used, whatever the specific numbers may be, involves writing down the following table of figures, as explained below:

12	17	12	29	46
2	11	1	7	2
	187	46	233	420
2	7	4	2	0
	1309	233	1542	2851
1	5	3	2	1
	6545	2851	9396	15941
3	4	3	0	1
	26180	2851	29031	55211
				etc.

We start, for convenience, with the larger columns and write down, on the first line, the remainder of 12 when the men are arranged after the battle in columns of 17. Obviously if this had been the only tally made, the count could have been 12, 29, or 46, etc., which we write after the 17 on the first line, adding 17 to each previous figure. The next requirement is for a remainder of 2 when there are 11 men in a squad; so in the next line we write the 2 and the 11 below the corresponding figures in the first line. Then mentally divide each of the previous answers by 11, setting down the remainders below each dividend until we come to the required remainder of 2, below the figure 46. On the next line we set down the product of 17 and 11, which is 187, and add that amount successively to 46, thus obtaining all counts which will meet both of the first two requirements. We now similarly apply the remainder of 2 for columns of 7, arriving at 233 for the smallest number satisfying the first three conditions, and continue with the remainder of 1 for columns of 5, and the remainder of 3 for columns of 4, obtaining all the possible final answers of 2851, 29031, 55211, etc., which meet the five conditions. Since it is likely from the given facts that the general lost 2,430 men rather than gained 23,750 or 49,930 or 76,110, etc., we confidently settle on the figure of 2851 men after the battle. Actually any answer equal to $2851 + 26180\,n$ is admissible and constitutes the full solution. It is to be noted that 26180 is the product of the figures 4, 5, 7, 11 and 17 for the five columns. Roy C. Weidler, Jr., of Los Alamos, New Mexico, contributed this solution.

37. THE SQUARE CORRAL

In tackling the apparently simple problem of laying out the largest possible square corral within a triangular plot, the tendency is to make two initial assumptions, both of which actually require proof (see below); first, that one side of the corral would be a side of the triangle, and second, that this side would be the shortest side. The simplest approach would be to do the job geometrically and then make the algebraic calculation, using the diagram (Fig. 118) for the purpose.

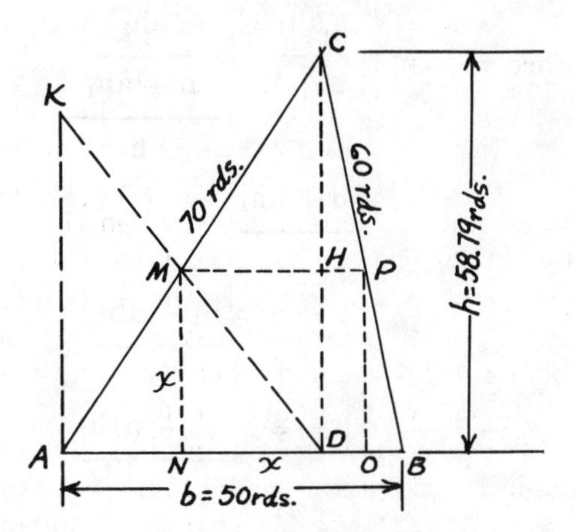

Fig. 118

Applying this method, CD and KA are drawn perpendicular to AB, KA being equal to AB. Line KD is joined and the required square MPON located by drawing MP parallel to AB, and MN and PO parallel to CD. For proof, by similar triangles and axiom one, MP/AB = MH/AD. Also MH/AD = ND/AD = MN/KA = MN/AB, from which MP must equal MN.

Using now the familiar formula for the altitude h of the original triangle in terms of the length of its sides, $h = (\sqrt{s(s-a)(s-b)(s-c)})/(a/2)$ we get h = 58.79. From similar triangles CMP and CAB we have MP/AB = CH/CD which converts to $x/50 = (58.79 - x)/58.79$, $x = 27$ rods. Applying the same procedure to the larger sides, CB and AC, would give a slightly smaller square which leads us to the question (see above) of whether there is a general proof that the shortest side should be selected.

Calling the shortest side a, we may write the equation $x_1/a = (h - x_1)/h = 1 - x_1/h$ which may be written $x_1 = ah_1/(a + h_1)$. By symmetry, for a square erected on a larger side b, $x_2 = \dfrac{bh_2}{b + h_2}$ but since $ah_1 = bh_2$

$$x_2 = \frac{ah_1/(b + ah_1)}{b}$$

$$\text{Therefore } x_1 > x_2 \text{ if } \frac{ah_1}{a + h_1} > \frac{ah_1}{\dfrac{b + ah_1}{b}}$$

$$\frac{b + ah_1}{b} > a + h_1$$

$$b - a > h_1 - \frac{ah_1}{b}$$

$$b - a > \frac{(b - a)h_1}{b}$$

$$\text{or if } 1 > h_1/b$$

But for an acute angled triangle $b > h_1$, so $x_1 > x_2$. Similarly for an obtuse angle triangle it may be shown that this is true only beyond certain relative values of a, b and c.

38. CAPTIVATING PROBLEM IN NAVIGATION

This innocent-looking problem is unique in that it not only calls for a knowledge of the properties of special curves — in this case the equiangular spiral — but also applies higher algebra and analytical geometry. Also, it demands for a complete solution, a knowledge of probability and triangulation methods. Like Euclid's Put-Put Problem (No.

28), The Easiest Throw (No. 18), and a few others in this book, it constitutes a brush-up course in mathematics, in itself.

A few of the answers received were on the light side; one contestant, for example, stated that since the waters covered by a fog bank are usually calm and a ship making 15 knots in calm water leaves a comparatively strong narrow wake visible for two or three hours, Captain A could catch B by merely following his wake, since the nearer he got, the stronger the wake.

Most readers, however, realized that the conditions responded nicely to the properties of a logarithmic (i.e., equiangular) spiral, since the distance a point moves along such a spiral bears a constant relation to the advance, or increase in length, of the radius vector. This means that if the initial position of ship B were taken as the pole of a spiral having a uniform angle between the radius vector and the tangent of 60°, the cosine of which is the ratio of the speed of the

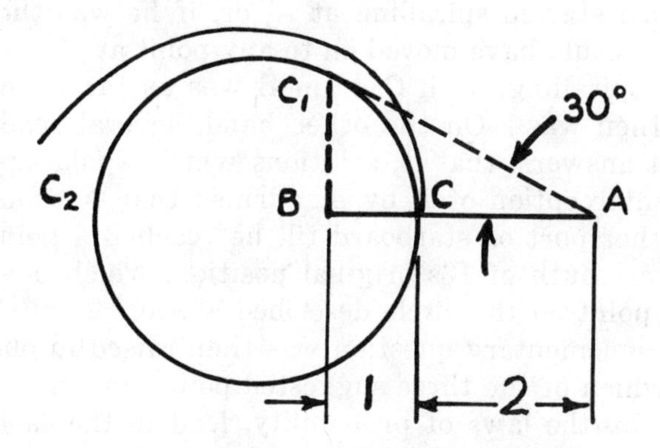

Fig. 119

two ships; Captain A could merely proceed to any point where he might conceivably intercept ship B and then follow such a spiral from there on. This, however, raises the question as to where such a point should be, since it is evident that the locus of such points is a circle of Apollonius (CC_1C_2 in Fig. 119) because it defines a point whose distance from one fixed point bears a given ratio — in this case 2 to 1 — to its distance from another fixed point. At this place, all DIAL readers missed the mark somewhat, since none attempted to prove whether or not there was any advantage in selecting any particular point, on such a circle or elsewhere, to start spiralling.

Most of the correct answers received specified that Captain A should continue towards the point where B entered the cloud bank for two thirds the original distance — point C in the diagram — between the ships. By this plan the spiralling obviously starts as soon as possible, which would seem to be an advantage. On the other hand Captain B cannot be expected to double back into clear view, so there is no chance of catching him at the spiralling point. Other readers affirmed with equal assurance that Captain A should continue an equal distance beyond B's initial position — up to point C_2 which delays the spiralling till the last moment. (Incidentally, A could also have made an abrupt stop and waited till B had gone the distance BC_2 in any direction, and then started spiralling at A, or, if he was the waiting kind, he could have moved on to any point at all and waited, before spiralling, until Captain B was as far from point B as A then was.) On the other hand, several readers with correct answers, that is, solutions which would assure ultimate interception of B by A, afirmed that A should swing 30° either port or starboard till he reached a point C_1 due north or south of B's original position, which obviously is also a point on the circle described above.

A supplementary question was then raised in our column as to which of the three suggested points to start spiralling would, by the laws of probability, lead to the earliest capture of the quarry. We quote an interesting answer to this question received from John Braden, Pako Corporation, Minneapolis: "From analytic geometry, the general equation of the spiral is $r = a\, e^{m\theta}$, where a is the value of r when $\theta = 0$ and m is cotangent of the angle between the radius vector and the curve. In this family of spirals $m = 1/3\sqrt{3}$, therefore the angle between the radius vector and the curve is

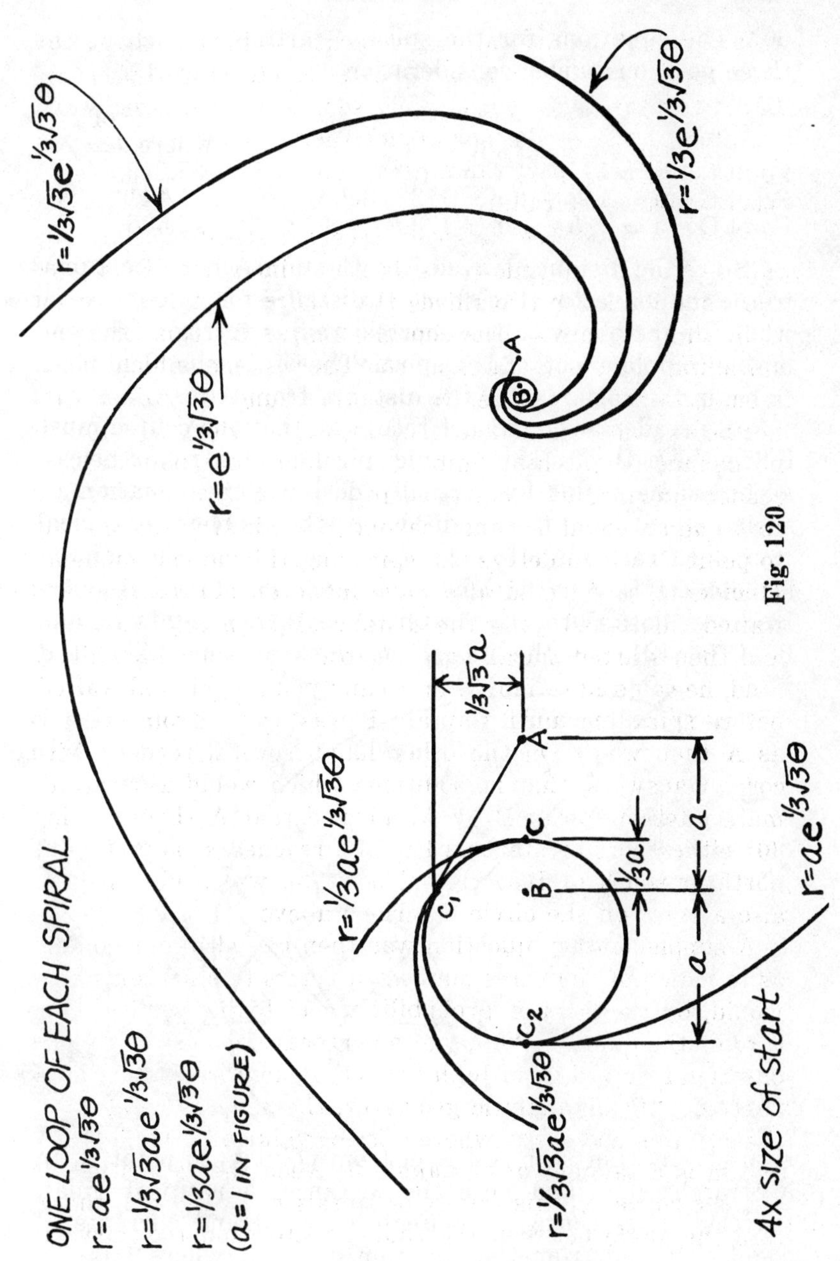
ONE LOOP OF EACH SPIRAL
r=ae 1/3√3θ
r=1/3√3ae 1/3√3θ
r=1/3ae 1/3√3θ
(a=1 IN FIGURE)
r=1/3√3e 1/3√3θ
r=e 1/3√3θ
r=1/3e 1/3√3θ
A
B
r=1/3ae 1/3√3θ
1/3√3a
A
C
B
C1
1/3a
a
a
C2
r=1/3√3ae 1/3√3θ
r=ae 1/3√3θ
4x SIZE of start

Fig. 120

60°. The equations for the spirals starting at each of the three positions under consideration are (see Fig. 120):

		Curve	value of r where $\theta = 2\pi$
Point C_2	$a = 1$	$r = e^{1/3\sqrt{3}\theta}$	37.619
Point C_1	$a = 1/3\sqrt{3}$	$r = 1/3\sqrt{3}e^{1/3\sqrt{3}\theta}$	21.141
Point C	$a = 1/3$	$r = 1/3\, e^{1/3\sqrt{3}\theta}$	12.536

"Since the distance traveled by Captain A is proportional to the radius vector, the shortest distance for a 360° search would be the one with the shortest radius vector at the end of one complete loop of the spiral. Therefore, the ideal place to begin the spiral is 2/3 the distance from A to B."

Some readers who did not recognize that the course must follow the path of a logarithmic spiral arrived, nevertheless, at the same result by virtually deriving the equation of such a spiral from the conditions of the problem. A typical course of reasoning follows (we quote from solution by Mr. Robert E. Smart, General Engineering Dept., Anchor-Hocking Glass Corp., Lancaster, Ohio): "Let points A_1 and B_1 (Fig. 121) be the initial positions of ships A and B

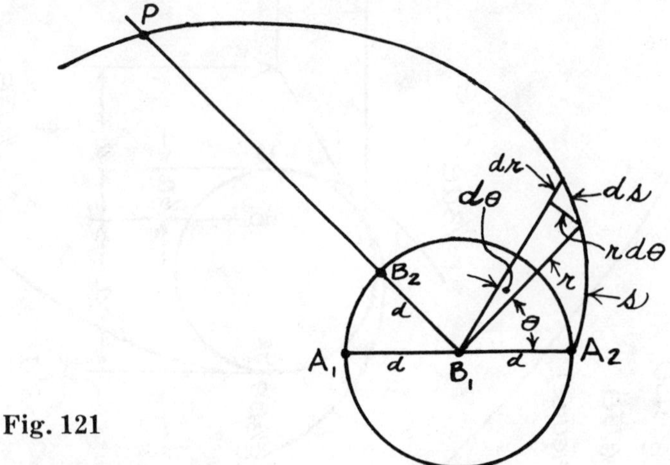

Fig. 121

respectively. Let ship A travel from point A_1 to point B_1 in a straight line, and note the distance, d nautical miles, traveled. (Neglecting the effects of wind and currents, d could be found from the equation $d = 30t$ where t is the time in hours required to travel the distance d in nautical miles). Let A continue to travel in a straight line for d nautical miles. (Actually the direction of this second leg of length d is immaterial). Now ship A has traveled 2d nautical miles, and in the same time ship B will have traveled d nautical miles. Both are now d nautical miles from point B_1 and therefore both ships lie on a circle with center B_1 and radius d. Ship A must now follow a spiral path, centering at point B_1, such that the magnitude of its radial component of velocity, $\frac{dr}{dt}$, is equal to that of ship B. And since B's velocity is entirely radial, $\frac{dr}{dt} = 15$. If this condition is met, both ships will always be the same distance from point B_1. To find the equation of this spiral path, let B_1 be the origin, and B_1A_2 the initial line of a polar frame of reference. In the infinitesimal triangle shown in the figure

$$(ds)^2 = (dr)^2 + (rd\theta)^2$$

But since
$$\frac{dr}{dt} = 15$$

and
$$\frac{ds}{dt} = 30$$

$$\frac{ds}{dt} = 2\frac{dr}{dt}$$

and
$$ds = 2\,dr$$

so
$$(2dr)^2 = (dr)^2 + (rd\theta)^2$$

$$3(dr)^2 = (rd\theta)^2$$

$$\sqrt{3}\,dr = rd\theta$$

$$\frac{dr}{r} = \frac{d\theta}{\sqrt{3}}$$

When
$$\theta = 0, r = d$$

$$\text{so} \quad \int_{d}^{r} \frac{dr}{r} = \int_{0}^{\theta} \frac{d\theta}{\sqrt{3}}$$

$$\Big[\log r\Big]_{d}^{r} = \Big[\frac{\theta}{\sqrt{3}}\Big]_{0}^{\theta}$$

$$\log \frac{r}{d} = \frac{\theta}{\sqrt{3}}$$

and $r = de^{\theta/\sqrt{3}}$ r in nautical miles, θ in radians.

"Thus, if Captain A follows the plan outlined, his ship will intercept ship B at some point P dependent upon the direction which ship B takes from point B_1"

39. THE SQUARE-DEAL SENATOR

This type of problem is familiar to the electrical engineer who uses it in calculation of networks — the details of which, however, are beyond the scope of this book. If all

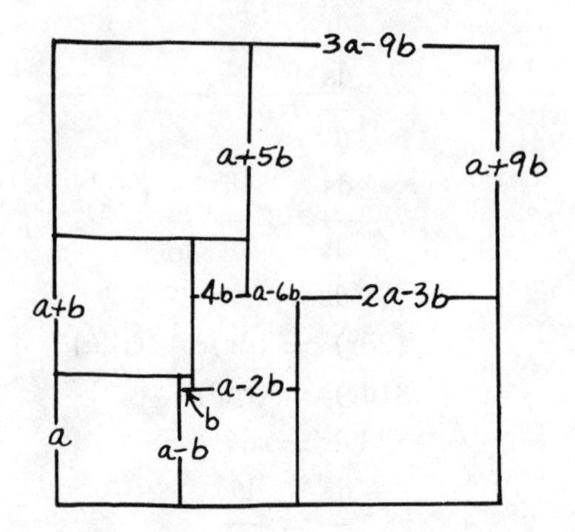

Fig. 122

9 squares are unlike, only two configurations are possible, and it may be shown also that a squared rectangle cannot

have less than 9 elements. In each case, two of the squares, as indicated (Figs. 122 and 123) may be assigned values of a and b, which establishes the corresponding values of all other squares and leads to the equation $a + 9b = 3a - 9b$ for the first case (upper right hand square) and $7a + 13b = 9a + 8b$ (total height of rectangle) for the second. Taking the smallest integral values of $b = 1$ for the first case, (a then equals 9) and 2 for the second (a then equals 5), the sides are respectively 1, 4, 7, 8, 9, 14, 15, 16, and 18; and 2, 5, 7, 9, 16, 25, 28, 33 and 36. Since the first set makes Euclid a definitely unwhiskered 9, whereas the second would make him 16 and in possible need of a razor, there is only one cor-

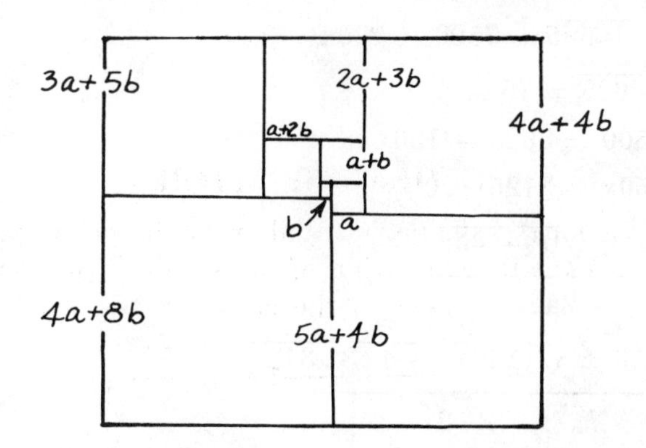

Fig. 123

rect answer (Fig. 122), which was calculated by this method by Martin Bressler, Armour Research Foundation, Chicago and Vincent Capurso, Piasecki Helicopter Corporation, Morton, Pa. Incidentally, the second configuration was reported to be newly discovered in an article on "Dissection of Rectangles into Squares" appearing in Duke Mathematical Journal, Dec. 1940. It is interesting, moreover, to note from the above that in order to split a rectangle into 9 unequal squares, the ratio between its height and base must be either 32/33 or 61/69 — a fact both mysterious and unpredictable.

40. THE MAST-TOP PROBLEM

To solve by conventional algebra (Fig. 124) an equation is set up which requires the succession of steps listed below:

Solution #1 (Algebra)

$$\sqrt{x^2 + 625} + \sqrt{(50 + x)^2 + 625} = 100$$

$$\sqrt{(50 + x)^2 + 625} = 100 - \sqrt{x^2 + 625}$$

$$(50 + x)^2 + 625 = 10,000 - 200\sqrt{x^2 + 625} + x^2 + 625$$

$$2500 + 100x + x^2 + 625 =$$
$$10000 - 200\sqrt{x^2 + 625 + x^2 + 625}$$

$$200\sqrt{x^2 + 625} = 7500 - 100x$$

$$2\sqrt{x^2 + 625} = 75 - x$$

$$4x^2 + 2500 = 5625 - 150x + x^2$$

$$3x^2 + 150x - 3125 = 0$$

$$x = \frac{-b \pm \sqrt{b^2 - 4ac}}{2a}$$

$$= \frac{-150 \pm \sqrt{(150)^2 - 4 . 3 . 3125}}{2 . 3}$$

$$= \frac{-150 \pm \sqrt{60000}}{6}$$

$$= \frac{-150 \pm 244.949}{6}$$

$$= \frac{94.949}{6}$$

$$= 15.824$$

A much simpler approach, however, is to consider the mast tops as foci of an ellipse, the foci for convenience being on the x axis and equally spaced from the y axis. This method is suggested from the definition of an ellipse, namely, the locus of a point so moving that the sum of its distances from two fixed points (the foci) is constant. The familiar equation of the ellipse as laid out above is $\frac{x^2}{a^2} + \frac{y^2}{b^2} = 1$, where in

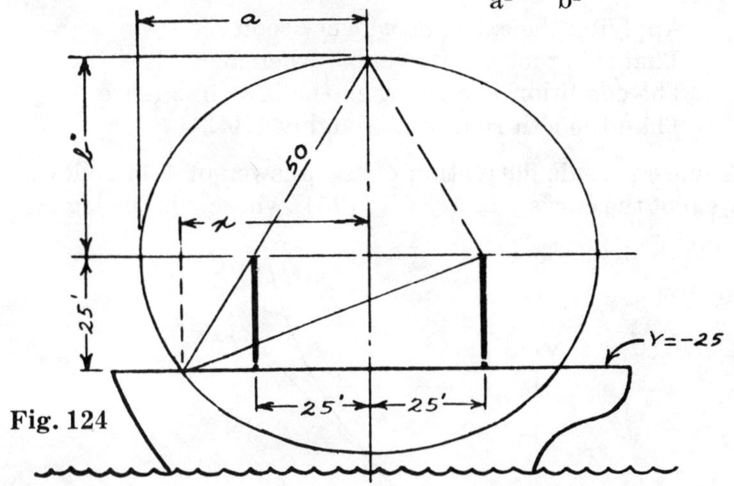

Fig. 124

this case a is 50, y is -25 and b is 43.3. This gives in very short order $X = 40.825$, the required distance being 25 less than that figure.

41. THE CHANGED BASE

When this problem appeared in the DIAL, it was incorrectly worded, due to typesetter's error, so as to read "Mr. Hendler observed that the logarithm of any positive number to that base is always less than the *base*," instead of the correct wording given in problem No. 41. This error was noticed by most readers; in fact Mr. Robert A. Harrington, B. F. Goodrich Research Center, Brecksville, Ohio, not only realized that the problem was somewhat "off base", but signalized this unprecedented event in verse:

Take a number whose log is $1 + b$,
Where b is the base, and it's plain to see
That the typesetter had a moment's slumber
When he set up "base" instead of "number."

If this is the case, how great must b be
For a to be less than the a'th power of b?
Here a is the log, and we can now say
That b must be more than the a'th root of a.

Applying the calculus, we very soon see
That this root has its "max" when a equals e.
The condition is met, when the base b is more
Than the e'th root of e: roughly 1.44.

A more prosaic derivation of the answer of 1.44 follows: If we plot the curve $y = a^x$ (Fig. 125), where x is the log and

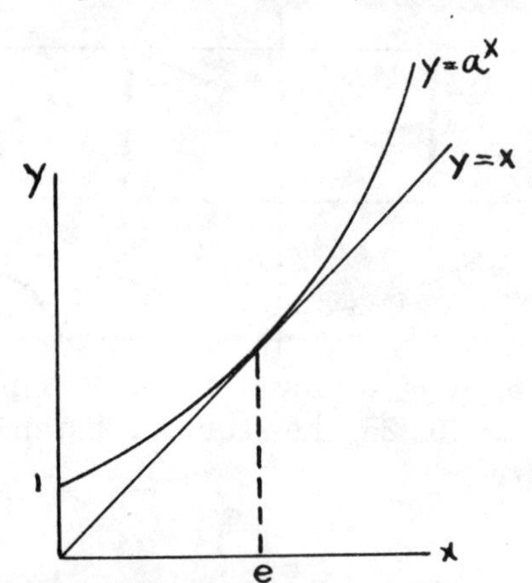

Fig. 125

y is the number, the problem is equivalent to finding the value of a which makes the curve $y = a^x$ tangent to the line $y = x$. For higher values of a, the curve lies entirely above $y = x$ and for lower values, the curve crosses $y = x$. At the point of tangency $dy/dx = a^x \log a = 1$ and $a^x = x$, from which $x = e$ and $a = e^{1/e} = 1.445$.

Another approach is to find the maximum value of a for $y = a^y$. Differentiating $dy = (ya^{y-1})(da) + (a^y \log a)\, dy$ from which $\frac{da}{dy} = \frac{1 - a^y \log a}{ya^{y-1}}$. Setting the derivative equal to 0, $1 - a^y \log a = 0$ and since $a^y = y$, $1 - y \log a = 0$ and $1 - \log y = 0$ from which $\log y = 1$ and $y = e$ and $a = e^{1/e}$ as before.

42. MATHMAN'S GREETING CARD

There are a vast number of ways to decipher the ingenious greeting card, but the simplest was contributed by D. J. White of Anglo-Canadian Pulp and Paper Mills, Ltd., of Quebec, Canada, who wrote: Starting from the left side, (first column) it is obvious that A is 0. C has to be one more than N since one is the maximum carry-over from the next column. H must be 4 or 9, which are the only single digit squares of an unlike digit, but it cannot be 9 because D and S plus carry-over could not then add up to 19. Therefore H is 4 and E is 2. The fourth column including carry-over must then equal 12, which means that D plus S in the third column equals 12, from which D and S are either 8 and 5 (in either order) or 7 and 6, since 4 has already been used. D cannot be 8 since then (last column) J and R would be the same numbers, nor can it be 5 or 7 since, from the fourth column, R would then have to be 4 or 2 respectively. Therefore, it must be 6 and R must be 3 and S 7. N must then be 8 and C 9, the only remaining consecutive numbers, which makes & 1 (fifth column) and J 5 (sixth column), with no solution other than the following:

$$\begin{array}{r} 6\;1\;5 \\ 0\;8\;6\;3\;2\;2 \\ 7\;2\;8\;6 \\ \hline 9\;4\;2\;2\;3 \end{array}$$

We suggest you compare this with your own solution, which may possibly be shorter or simpler.

43. MEETING OF THE PLOWS

Please refer to the simpler problem No. 12 about a snow plow that started to clear a street-car track at noon and got through a mile during the first hour, and a half mile in the second hour. The question was — when did the snow start to fall? You may recall that dozens of readers of the DIAL came through with an answer of 11:30 A.M., whereas the use of calculus, which unfortunately is a must in this type of problem, proved the correct answer to be 11:23 A.M. Strange to relate, the snow *did* start falling at 11:30 A.M. in this problem and we were reminded of this unusual coincidence by several readers who solved the problem correctly as below:

Let n denote the number of hours before noon that it started to snow and let x, y, z denote the distances that the first, second and third snow plow traveled in t hours past noon. Then $dt/dx = t + n$, $x = 0$ when $t = 0$, with solution $t = e^x n - n$. Also $dt/dy = t - (e^y n - n)$, $y = 0$ when $t = 1$, with solution $t = e^y(n + 1 - ny) - n$. Also $dt/dz = t - [e^z(n + 1 - nz) - n]$, $z = 0$ when $t = 2$, with solution $t = e^z(n + 2 - nz - z + nz^2/2) - n$. Let d denote the common value of x, y, and z when the plows meet and let T denote the value of t when the meeting occurs. Then $(T + n)/e^d = n = n + 1 - nd = n + 2 - nd - d + nd^2/2$. Solving we get $n = 1/2$ and $T = 3.195$. Therefore it began to snow at 11.30 and the plows met at about 3:12.

44. THE DIVIDED CHECKERBOARD

Mr. Grossman's checkerboard problem proved even more intriguing than expected. Instead of a mere exercise in cut-and-try, it turned out to be capable of verifiable analytic solution.

In the wide variety of solutions received the suggested number of cuts varied from a minimum of 8 to a maximum of 179; the latter included cuts that had been definitely ruled out in the statement of the problem. Favorite answers were 51, 65 and 73, and the top total of 95, shown in the drawing was submitted by three readers; Mr. Eric Schauer of Naval Ordnance Test Station, Pasadena, Mr. Duane R. Chicoine of A. E. Staley Mfg. Co., Decatur, Illinois, and Mr. Walter J. Dash, Kipling, Sask., Canada. Mr. Samuel M. Sherman, Naval Air Development Center, Johnsville, Pa., explained his method of analysis in ingenious fashion but unfortunately missed 10 of the possible cut-ups.

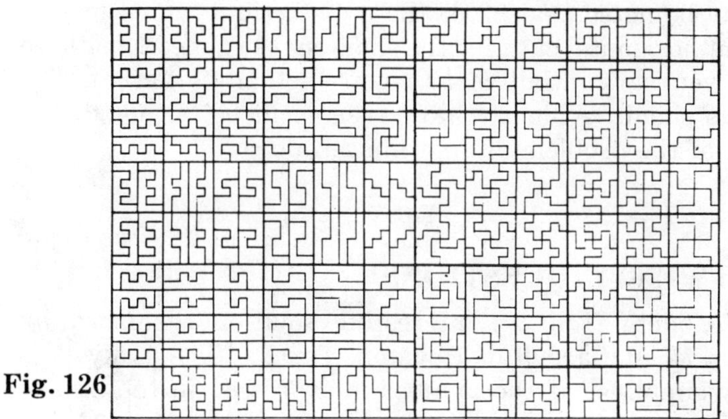

Fig. 126

Mr. Sherman correctly said that the cuttings could be considered either as 90° cuts, each piece being rotated 90° to lie on an identical piece, or as 180° cuts, in which the board was first cut in two parts (by either a horizontal or a vertical line) and then each half divided into two pieces, one of which would lie on the other when rotated 180°. To illustrate the first half of Mr. Sherman's method, we show two of each

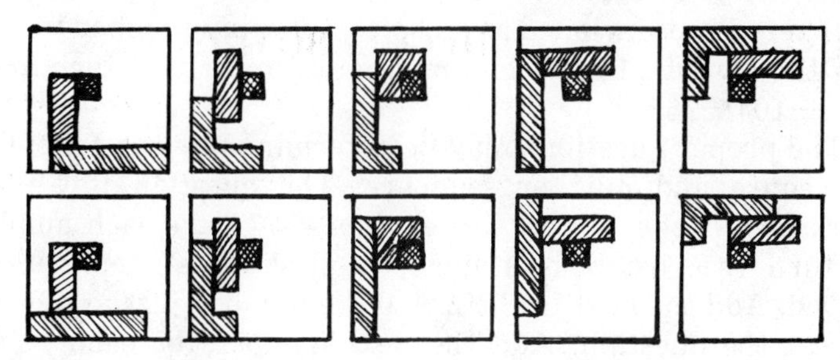

Fig. 127

of five different groups of 90° cuts (Fig. 127); there are a total of nine different cuttings in each group. Note that the single inner piece is always the northwest square of the 4 center squares, the middle section is made up of 3 squares and the outer section 5 squares. Mr. Sherman, however, missed 6 cuts in which the middle section is disjointed though firmly joined to the single inner square and the outer section. The full quota of 95 solutions is shown in Fig. 126. (It may be noted that 22 of these dissections which have a straight horizontal cut give the same *shape* of pieces as those with a vertical cut, but that they are actually different. The cuttings should evidently have been included, although the author of the problem did not himself do so.)

45. TWENTY QUESTIONS

We quote from an interesting solution furnished by William M. McCardell of Houston, Texas: "The analysis is greatly simplified if the number is expressed in the binary scale. Since there can be only two possibilities (0 or 1) for each digit, the value of each digit is uniquely determined by a single reply. Obviously, then, the largest number which may be determined by n questions is the largest n digit binary number (if zero be included as a possible choice):

For $n = 20$, this number is: 11111111111111111111

The value of this number in the ordinary (denary) scale is $2^{20} - 1 = 1048575$.

If zero is not included as a choice, then any number including the lowest 21-digit binary number may be determined: $2^{20} = 1048576$.

The proper questions may be determined by the following procedure and number sequence. The question is always phrased, "Is the number larger than — ?" Take each number in turn (see list below) until the first "Yes" reply is received. Add the next in the list to the number just asked and repeat the question. Add the next number for each "Yes" answer, subtract the next number for each "No" answer until only two possibilities remain. One more question determines the number. Twenty questions will be required.

If zero is not a choice: $2^{19}, 2^{18} \ldots\ldots 2^{1}, 1$.

If zero is a choice, subtract 1 from each number:

$2^{19} - 1, 2^{18} - 1 \ldots\ldots 2^{1} - 1, 0$."

The same answer can, of course, be arrived at by process

of reasoning without reference to the binary system, and this was done by many readers who merely showed how the number of possibilities could be reduced by one-half with each question, but this is not nearly so neat an approach, nor does it lend itself so readily to consideration of the effect of including zero as a possible number. Incidentally, for those who may be unfamiliar with the binary and other possible systems of numeration, we might say in passing that our present denary system is doubtless derived from the fact that we have ten fingers. As we all know, we use digits from one to nine and move to the second place when we reach 10^1, to the third place when we reach 10^2, to the fourth place when we reach 10^3, etc. Any other base than 10 could, of course, be used in the same way and in fact, there are decided advantages to a senary, or six system. With our denary system, for example, three equal partners, which is a normal sized group, just cannot go into business and put up a total amount of $10,000 which is a good round sum. If we had a senary system, however, this would be simple. The round sum of $10,000 would, of course, be considerably devalued as compared to our present system, but each of the three partners could then subscribe the equally round sum of $2,000 and all would be serene.

Although the senary system does not find particular application in mathematical problems, the binary system does have highly important modern uses, particularly in electronic computing machines. Here it is used instead of ten because the binary system identifies every number, no matter how large, by a mere "yes" or "no" at every place — a most convenient two-state instead of ten-state type of machine operation.

For other examples of the binary system, see problems No. 30 and No. 70 on the game of Nim and the Triangle of Coins and, in lighter vein, mathematical nursery rhyme on Little Bo-Peep, opposite page 11.

46. CHILDREN AT PLAY

Wrote Virginia F. Brasington, American Enka Corporation, Enka, North Carolina, "The maximum number of children possible is seventeen. Under this condition, several house numbers exist from 24 to 140, which are the product of the children of four families decreasing in size to a minimum of 1; and several house numbers from 120 to 240, which are the product of the children of four families decreasing to a minimum of 2. The maximum number in the smallest family is 2, because if higher than 2, the total for four families would be a minimum of eighteen. When the question was asked, 'Does your cousin's family consist of a single child?', it was known that the house number was 120, because that is the only house number in which the minimum number in the family may be either 1 or 2. However, the combination of 8, 5, 3 and 1 or 6, 5, 4 and 1 can equal 120, whereas the only combination including 2 is 5, 4, 3 and 2. Therefore, the answer to the question must have been, 'no' and the vistor knew that the numbers of children in the respective families were 5, 4, 3, and 2."

47. THE FREE PLOT

Clearly, such a plot must give a maximum ratio of area, expressed in square miles, divided by perimeter expressed in miles. Several of our readers were familiar with the readily-proved relation that a given length of fencing will enclose the greatest area if it is shaped around a circle, so they took the problem as a Quickie and picked for Farmer Brown's plot the inscribed circle whose ratio of area over perimeter is, of course, $\pi/4$ divided by π or .25. Several of these readers took time to observe, however, that the same ratio would be had if Brown took the whole ranch, namely, 1 divided by 4, or .25. C. E. Norton, National Malleable & Steel Castings Co., Cicero, realized that a better deal would be had if Farmer Brown purchased the entire mile square, except that he rounded the corners to radius R. (It may be shown that such basic shape of plot must give the maximum ratio). It remains then to find R (see Fig. 128) such that Area/(length of fence) is a maximum. The equation is

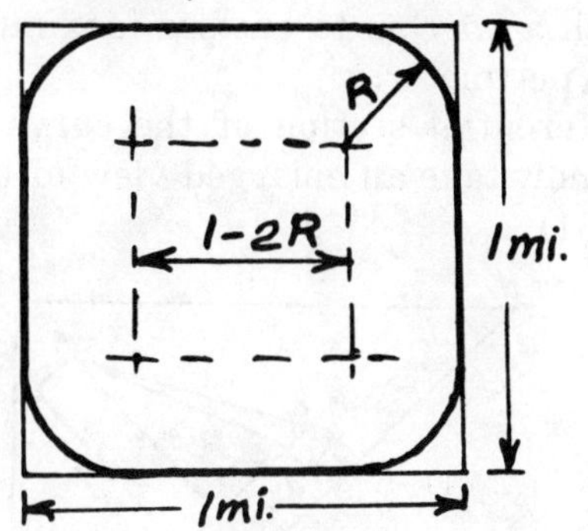

Fig. 128

readily written $A/L = (1 - 4R^2 + \pi R^2) / 4(1 - 2R) + 2\pi R$. Setting $d(A L)/dR$ equal to zero, R is readily found to be $1 / (2 + \sqrt{\pi})$ or .265 miles. The ratio of area to perimeter is that identical figure, namely R, whose identity too, is explained in author Salmon's more complete solution given below. This shows, for those who want to extract its full essence, that this innocent-looking problem ,unlike most in our collection, requires a really involved analysis for a fully correct solution.

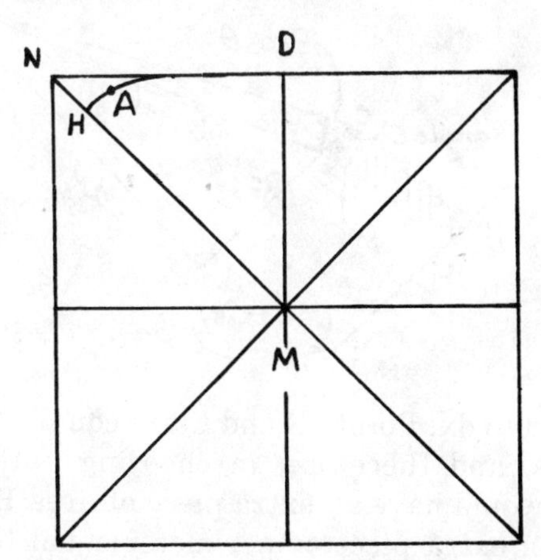

Fig. 129

Divide the square into eight sections (Fig. 129): By symmetry, we need consider only one of these. Assume DH represents the required curve. Point A is a point on the curve. At any point, such as A, the problem is to get to the MN axis in such a way as to enclose maximum area while using minimum perimeter.

Taking a differential section of the curve from A to B (Fig. 130), we now take an enlarged view of the differential section (Fig. 131).

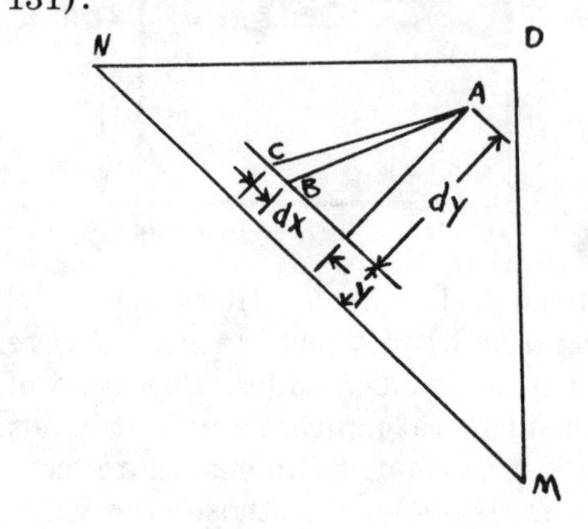

Fig. 130

Assume the most favorable path is from A to B. Now consider the alternate path AC. The distance P is infinitesi-

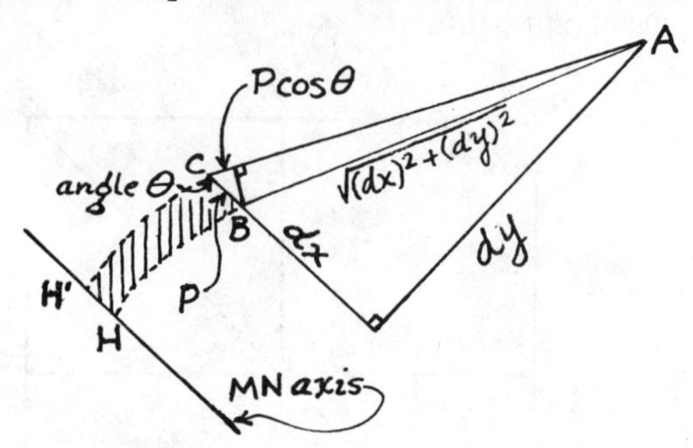

Fig. 131

mal compared to dx. Points B and C are equally distant from the MN axis, and therefore, in choosing path AC rather than AB we would have an extra piece of area BCH′H at the expense of an extra piece of perimeter equal to P cos θ. If now BH is the most favorable path from B to the MN axis, then CH′ must be parallel to BH. The area of BCH′H, then, is Py. The added perimeter P cos θ is expressed in terms of x and y as follows:

$$\cos \theta = \frac{dx}{\sqrt{(dx)^2 + (dy)^2}} = \frac{1}{\sqrt{1 + \left(\frac{dy}{dx}\right)^2}}$$

$$\text{added perimeter} = P \cos \theta = \frac{P}{\sqrt{1 + \left(\frac{dy}{dx}\right)^2}}$$

Now it will be shown that the ratio of added area (Py) to added perimeter must be equal to the ratio of area to perimeter for the whole figure. (Call the ratio R.) For if AB is the most favorable path, then any deviation from this path must result in adding such amounts of area and perimeter as to diminish the overall ratio R. The only way this can be done is by having the ratio of the added amounts be less than R. But since AC is only a differential amount removed from AB, the ratio of the added amounts approaches R as a limit as the path approaches AB. So we have now the ratio of added area to added perimeter equal to R, or:

$$\frac{Py}{\frac{P}{\sqrt{1 + \left(\frac{dy}{dx}\right)^2}}} = R, \text{ or } dx = \sqrt{\frac{y^2}{R^2 - y^2}}\, dy$$

Integrating, we have: $x^2 + y^2 = R^2$, the equation of a circle of radius R. Now we have to locate its center. It can be seen that in starting out from point D the curve must follow the edge of the square for some distance. This is because any deviation from the edge of the square must be compared with the alternate path of following the edge of the square using the test described above for paths AB and AC. As formerly, the ratio of added area to added perimeter (by choosing the edge of the square rather than deviating from it) is:

$$\frac{Py}{\frac{P}{\sqrt{1 + \left(\frac{dy}{dx}\right)^2}}} \text{ or } y \sqrt{1 + \left(\frac{dy}{dx}\right)^2}$$

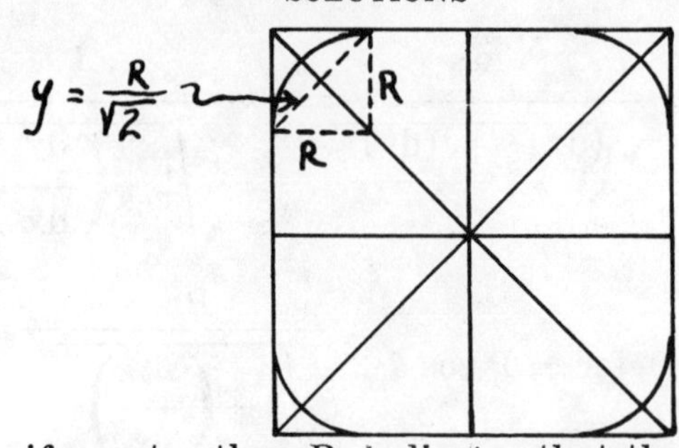

Fig. 132

This ratio, if greater than R, indicates that the edge of the square is the more favorable path. At point D, $y = \frac{a}{2\sqrt{2}}$ (where a is one side of the square) and $dy/dx = 1$ (using MN as the x axis). So $y\sqrt{1 + (dy/dx)^2} = a/2$ at point D. This is greater than R can ever be, because even if we took as the smallest possible perimeter a perpendicular from D to MN, this would give a perimeter of $\frac{4a}{\sqrt{2}}$ for the whole figure. And even if we took the whole area of the square (a^2) as the area, the ratio R would be $\frac{a^2}{4a/\sqrt{2}}$ or $\frac{\sqrt{2}\,a}{4}$ or .353 a. So at point D it follows that the edge of the square is the most favorable path. This condition persists until $y\sqrt{1 + (dy/dx)^2}$ is equal to R, which occurs when $dy/dx = 1$ and $y = \frac{R}{\sqrt{2}}$. Beyond this point, deviation from the edge of the square is more favorable than following it. It follows that at this point the circle is tangent to the edge of the square, and this locates the center of the circle and defines the whole figure (Fig. 132). This is a square with circular fillets at each corner with radius R.

The next problem is to evaluate R. The area of the entire figure is $a^2 - 4(1 - \frac{\pi}{4})\,R^2$. Its perimeter is $4(a - 2R) + 2\pi R$. The ratio of area to perimeter is R, or

$$R = \frac{a^2 - 4(1 - \pi/4)\,R^2}{4(a - 2R) + 2\pi R}$$

From which $R/a = 1/\ (2 + \sqrt{\pi}) = .265$ or: $R = .265$ a.

It might be noted that the square itself has a ratio R = .250 a. This is also true of a circle inscribed in the given square. So the maximum possible ratio of area to perimeter under the conditions given is .265.

48. THE BALANCED PLANK

A simple solution to this interesting problem is contributed by Louis A. Feathers, ANPD, General Electric Co., Cincinnati: "If the plank is to have neutral equilibrium, its center of gravity in any displaced position must be directly over the point of tangency on the cylinder. Since the plank is displaced by rolling, the distance from the point of tangency to the top point of the cylinder is the same as the dis-

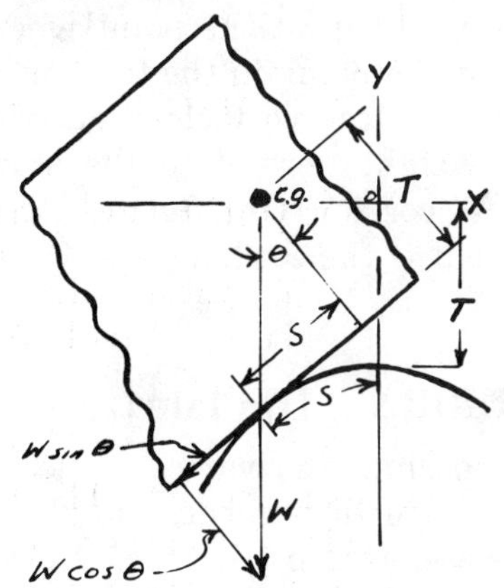

Fig. 133

tance along the plank from the point of tangency to the point on the surface which is in the normal plane through the c.g. The following equations can then be written:

$$\frac{dy}{dx} = -\frac{S}{T}$$

$$-T\frac{d^2y}{dx^2}=\frac{ds}{dx}$$

$$\text{and}\ \frac{ds}{dx}=\sqrt{\left(\frac{dy}{dx}\right)^2+1}$$

$$\text{then}\ \frac{d^2y}{dx^2}=-\frac{1}{T}\sqrt{\left(\frac{dy}{dx}\right)^2+1}$$

$$\frac{dy}{dx}=-\sin h\frac{x}{T}$$

$$y=-T\cos h\frac{x}{T}$$

Thus the cylinder is shaped like an inverted catenary, since the familiar curve in which a uniform cable hangs when suspended from two fixed points has that equation. The plank will begin to slip at the point where the component of the weight of the plank in the tangential direction is equal to the product of the coefficient of friction and the component of the weight normal to the surface. At that point $W \sin \theta = f\, W \cos \theta$, from which the critical angle is the one whose tangent is the coefficient of friction.

49. THE RECURRING DECIMAL

Of the many possible and varied modes of solution to this interesting problem in number, the shortest was submitted by H. Weisselberg of Leonia, New Jersey, who wrote:

"We start with the determination of the divisor. It has to contain for a nine-place recurring decimal a factor of 999,999,999 $= 81 \times 37 \times 333{,}667$ and also, because of the non-recurring one decimal, the factor 2 or 5. Since it has to be a number of six figures, the only solution for the divisor is $2 \times 333{,}667 = 667{,}334$.

The third remainder and also the last, after attaching zero, must be at least 1,000,000 (7 figures) and at most 1,334,660 (smaller than $2 \times 667{,}334 - 1$ and ending with zero). So its second figure can be only 1, 2, or 3. The 8th remainder after attaching 0 has the form

```
x y 0 0 0 0  Subtracting
6 6 7 3 3 4  (divisor, only 6 figures)
-----------
1 z 2 6 6 6  where z from above can be only 0, 1, 2 or 3
```

```
667334)7752341(11.6168830001
       667334
       1079001
        667334
       4116670
       4004004
        1126660
         667334
        4593260
        4004004
         5892560
         5338672
          5538880
          5338672
           2002080
           2002002
                780000
                667334
                112666
```

Fig. 134

It follows that xy can be only 77, 78, 79 or 80. y has to be even (difference between two even numbers) and therefore 77 and 79 have to be discarded. 80 is also not possible because:

We would have

```
--------0
3336670  5 × 667,334, the only one to have
-------  a zero as last figure.
     80
```

This would mean that the 7th remainder would have an uneven last figure, which is not possible.

The only possible solution is therefore 78. This gives for z the value 1 and the third and 12th remainders are 112,666. The only multiple of 667,334 to have 7 figures and to add up to 112,666 to give a zero as last figure is $4{,}004{,}004 = 6 \times$

667,334. Based on these numbers after figuring back we obtain the solution as shown in (Fig. 134).

Another solution offered by Robert L. Patton, Engineering Department, Gulf Oil Corp., Pittsburgh, instead of using the above analysis went back to first principles governing recurring decimals and got the answer without finding the figure 78. Any fraction which reduces to an n-place recurring decimal, must equal the recurring decimal itself divided by n nines. Thus the recurring decimal .142857 must be obtained from the fraction 142857/999999 or its equivalents 1/7, or 2/14 etc. The third and last remainders in this long division must therefore equal the 9-place recurring decimal in the quotient, divided by 2997 and multiplied by 2 because 999,999,999 divided by 667,334 gives 1498.5. But the first and last digits of the 9-place repeating decimal must each be 1, since no other multiple of 667334 produces only 6 digits. "Therefore," writes Mr. Patton, "we look for a number A, such that $2997 \times A = 1 \ldots . 001$," and it takes only a few minutes to discover that A must be 56333, which is multiplied by 2 to give the third and last remainders, the rest of the figures then being readily filled in.

50. THE HUNTER AND HIS DOG

When this pursuit problem appeared in the DIAL, we were struck by the great variety of solutions received; no two correct answers used the same method of approach. A favorite solution assumed that the dog was smart enough to pick a straight line course to the nearest possible meeting point. On this basis, by the time dog met man the portion of the dog's speed of 3 mph that was parallel to the stream must have overcome the man's speed of 1 mph plus the stream flow of 1 mph or a total of 2 mph. Since the portion of the dog's speed that was perpendicular to the stream would be $\sqrt{3^2 - 2^2}$ or $\sqrt{5}$, this figure times the time would then equal the stream width of one mile, from which the time is $1/\sqrt{5}$ hrs. and the hunters' travel would be the same distance in miles or .447 miles. Unfortunately, this convenient solution ignores the statement in the problem that the dog, as expected, would keep swimming toward his master and thus follow a curved rather than a straight line path. This would of course give him a longer swim, — the penalty for being a dog. The problem which thus deals with small functional changes of direction is not capable of solution except by the "mathematics of small functional changes" commonly called the calculus, although most readers who used the calculus made the job quite involved and in some cases even used handbook formulas for pursuit curves. Weston Meyer of Bundy Tubing Company, Research and Development Laboratories, Birmingham, Michigan, did a really neat trick with equations that went back to the basic principles of the calculus and made the task look quite simple. Wrote Mr. Meyer: "Clearly the problem remains unchanged in essence if we suppose the stream to be at rest and add its rate of motion to that of the hunter. Thus, in

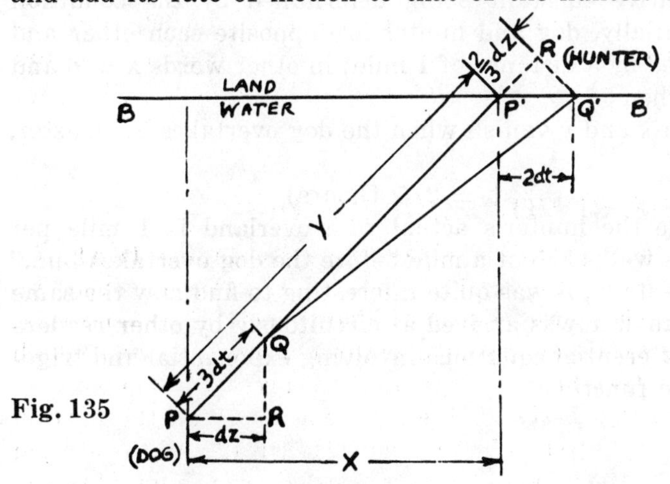

Fig. 135

Fig. 135, the water is motionless and the hunter moves along the bank to the right with a speed of 2 miles per hour. When he is at P′ the dog is at P and swimming 3 miles per hour in the direction of PP′. The length of PP′ is denoted by y and its projection on BB by x. In an infinitesimal interval of time dt (hours) the hunter moves from P′ to Q′, a distance 2dt, while the dog moves from P to Q, a distance 3dt. Lines PR and QR are drawn, parallel and perpendicular, respectively, to BB, to form triangle PQR. PP′ is extended to R′ where it meets the perpendicular from Q′ to complete triangle P′Q′R′. Triangles P′Q′R′ and PQR are evidently similar, and their corresponding sides in the ratio of 2 to 3. If PR = dz, then P′R′ = 2/3dz, as shown.

"In the interval dt, x is increased by 2 dt and decreased by dz, while y is increased by 2/3dz and decreased by 3 dt. Thus:

$$dx = 2dt - dz$$
$$dy = 2/3dz - 3dt$$

Eliminating dz:

$$dy + 2/3dx = -5/3dt$$

Integration is immediate:

$$1 - y - 2/3x = 5/3t$$

the additive constant being determined by the condition that, initially, dog and hunter are opposite each other and separated by a distance of 1 mile; in other words x = o and y = 1 when t = o.

"Both x and y vanish when the dog overtakes his master, so:

$$t = 3/5 \text{ (hours)}$$

"Since the hunter's actual rate overland is 1 mile per hour, he walks 3/5 of a mile before the dog overtakes him."

Incidentally, it was quite interesting to find how the same simple answer was arrived at circuitously by other readers using differential equations involving exponential and trigonometric functions.

51. RUSSIAN MULTIPLICATION

Little Euclid's feat of multiplying 85 x 76 (or any other pair of numbers) by use only of addition, multiplication and division by two, attracted an unusually wide variety of explanations. It encouraged many reminiscences on the part of our readers, several of whom reminded us that this is the so-called Russian method of multiplication, used by peasants in many sections of Russia before the advent of the Iron Curtain. Nearly all readers solved the problem by successively dividing the smaller number by 2 (ignoring the remainder, if any) and successively multiplying the larger number by 2 and then adding only those partial products which were alongside odd quotients, as below:

~~76x~~	~~85~~
~~38~~	~~170~~
19	340
9	680
~~4~~	~~1360~~
~~2~~	~~2720~~
1	5440
	6460

This method, while quite correct as to the "how," does not give the key to the "why" as well as if the multiplier is set above the first column and the remainder after each division is set in an intermediate column, as below:

76		
~~38~~	~~0~~	~~85~~
~~19~~	~~0~~	~~170~~
9	1	340
4	1	680
~~2~~	~~0~~	~~1360~~
~~1~~	~~0~~	~~2720~~
0	1	5440
	0	6460

In this arrangement only those products are added which are alongside a remainder, and in reality each product is multiplied by the remainder (which in this special case is, of course, either 0 or unity) and the products then total to give the answer.

This method of multiplication is not limited in any way to the binary system (sad to state, not a single reader brought out this fact) but is based on the familiar relation that any number may be expressed as the sum of powers of a smaller number, a digit for example — by successively dividing the number by the digit and setting off the remainders, which, when multiplied by successive powers of the digit, starting with zero and increasing by one with each division, gives products that add up to give the original number. This, of course, is what we do unthinkingly with any number when we express it in our decimal system.

For instance, if the number is 138 and the digit is 5, write as below: (R stands for remainder, and P for power)

138	R	P	R x P
27	3	1	3
5	2	5	10
1	0	25	0
0	1	125	125
			138

(If we were using a quinary system number 138 would, of course be written 1023 — the remainders in inverse order.)

Clearly, if 138 is to multiply any number whatever, such as 151, it would be necessary merely to multiply that number successively by 5, multiply each product by the adjacent remainder found as above (the original number being multiplied by the first remainder) and add, thus:

138			
27	3	151	453
5	2	755	1510
1	0	3775	0
0	1	18875	18875
			20838

With this explanation, which is credited in large part to Samuel M. Sherman of the Aviation Armament Laboratory,

Naval Air Development Center, Johnsville, Pa., it becomes quite simple to show how Euclid could have done the same type of job, but not quite so handily, if his studies had taken him as far as 3 in his multiplication and division tables:

76			
25	1	85	85
8	1	255	255
2	2	765	1530
0	2	2295	4590
			6460

52. THE INSIDE TRIANGLE

This problem, when it appeared in the DIAL, proved to be subject to more different modes of handling than perhaps any other in our entire seventeen-year collection. Interesting and accurate solutions were had by vector analysis, analytical geometry, and by straight geometry using a wide variety of construction lines to shorten the job. The simplest solution made use of the obvious fact that since the nature of

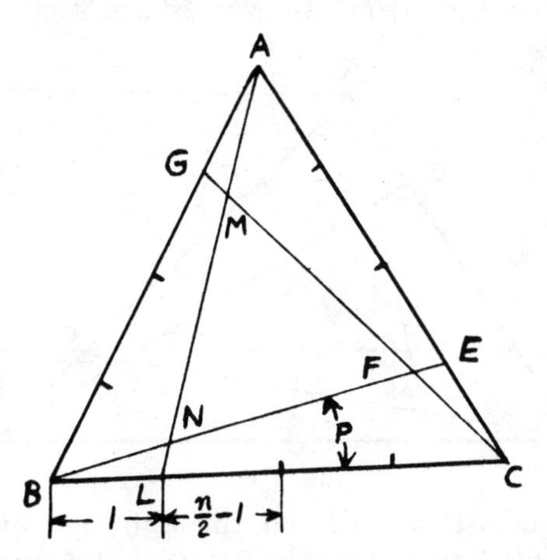

Fig. 136

the triangle was not specified, the ratio of areas expressed as a function of the independent variable n must apply to any triangle and, therefore, to the simplest kind, an equilateral triangle, which is also clear from topological considerations. This is basically an approach by symmetry.

Using the notation where n represents any division, and the section of side struck off is taken for convenience as unity, it is evident that

$$AL = \sqrt{(n\sqrt{3}/2)^2 + (n/2 - 1)^2}$$

or $\sqrt{n^2 - n + 1}$ which for convenience may be called k. Applying the law of sines to triangle ABL (and noting that by symmetry all corresponding triangles are equal and the central triangle MNF is equilateral), we have sin p = sin 60/k. Also, from triangle ANB, NB = sin p/ sin 60 or n /k. Similarly, from triangle NBL, NL = sin p /sin 60 or 1 /k. MN = AL — (AM + NL) which from the above reduces to $(k^2 - (1 + n))/k$. Remembering that the required ratio of areas is equal to $(MN)^2/n^2$ we obtain the answer $(n - 2)^2 / (n^2 - n + 1)$ which when n equals 4, reduces to 4/13. This solution was contributed by Walter Marvin, Jr., Raytheon Mfg. Co., Waltham, Mass., and by R. M. Brown, National Forge & Ordnance Co., Irvine, Pa. W. A. Blaser, Davenport, Iowa, used the simplest construction lines to achieve a like result by similar triangles without resort to the equilateral triangle (see Fig. 137). In Mr. Blaser's solution, parallel

Fig. 137

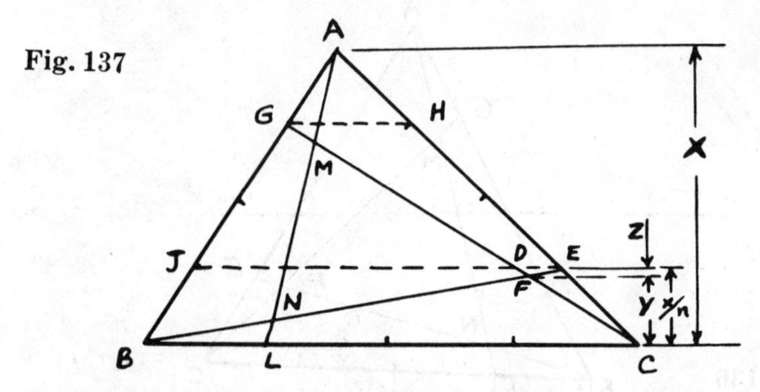

lines GH and JE are drawn through the division points. Then GH = BC /n and DE = GH /n — 1 from which DE = BC /n (n — 1) or BC /DE = n(n — 1). Since DEF and

BFC are similar triangles, $y/z = BC / DE = n(n-1)$ or $(y+z)/y = n(n-1)+1)/(n-1)$. Since $y + z = x/n$, $x/yn = (n(n-1)+1)/n(n-1)$ or $x/y = (n(n-1)+1)/(n-1)$. Since they have identical bases $BFC/ABC = y/x = (n-1)/(n(n-1)+1)$ or $BFC = ABC\ (n-1)/(n(n-1)+1)$. In similar fashion triangles MAC and ANB are readily shown to be equal to the same expression, and, since the area of triangle NMF equals that of the large triangle less three times that amount, the required ratio in areas is found, as above.

Perhaps the simplest solution of all was contributed by Howard D. Grossman of New York who wrote, "From the statement of the problem we may assume that these lines divide any triangle whatsoever into the SAME proportions. Then we can use the symmetry that would exist in an equi-

Fig. 138

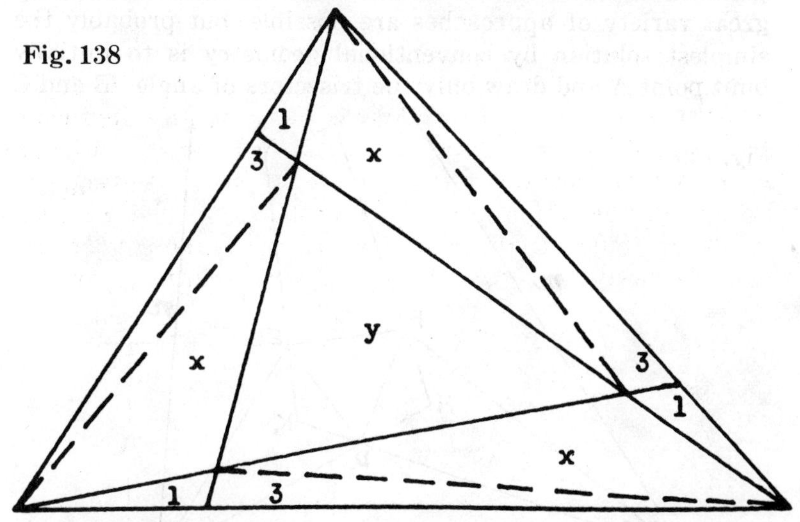

lateral triangle. Draw dotted lines as in Fig. 138. We use the proposition that triangles with the same altitude are in proportion to their bases. Call each of the three smallest triangles the unit of area. Then each of the triangles marked 3 has three times the unit area. Of the four remaining triangles, mark the central one y and the others x. Then by the above proposition, $x + y + 3 = 3(x + 1)$ and $2x + y + 7 = 3(x + 5)$, when $x = 8$, $y = 16$. Since the whole triangle $= 52$, the central one $= 4/13$ of the original. (By comparing triangles, we also find that each line from a vertex to the division point of the opposite side, is divided in the ratio 4:8:1.) By substituting n for 4 in this solution, we find similarly that the central triangle $= (n - 2)^2/(n^2 - n + 1)$ of the original (and each internal line from a vertex is divided into the ratio $n:n(n - 2):1$)."

53. TRISECTORS

The solution of this problem requires the adroit use of the most effective and time-saving construction lines. A great variety of approaches are possible, but probably the simplest solution by conventional geometry is to initially. omit point A and draw only the trisectors of angles B and C

Fig. 139

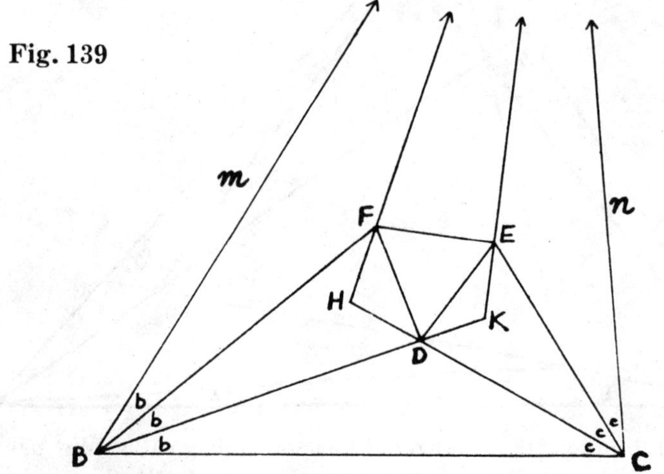

(Fig. 139). Since A when determined will be the third vertex of triangle ABC, $a + b + c = 60°$. Draw DF and FH so that angle HDF = angle HFD $= a + c$.
Draw DE and EK so that angle KDE = angle KED $= a + b$.
Therefore angle FDE $= 360° - (a + 120) - 2(b + c) - (a + b) - (a + c) = 60°$.

Angle BFD = 180° − b − (a + c) − (b + c) = a + 60 = 180° − c − (b + c) − (a + b) = angle CED. D is point of intersection of angle-bisectors of the triangle whose sides are BC, BF...._, CE...._, and so is equidistant from BF and CE. Since DF and DE make equal angles with BF and CE respectively, DF = DE and triangle DEF is equilateral.

It remains to prove that m, HF, KE, and n converge to a point where they form three equal angles. Since both small triangles on base DE are isosceles, KF bisects angle K, BF and KF are angle bisectors, and F is the point on third angle bisector of triangle m-BK-KE. Angle BFH = a + 60° − (a + c) = a + b. Angle m-KE = 180° -- 2b − (180° − 2a − 2b) = 2a. Therefore HF is the bisector of angle m-KE. Similarly KE is the bisector of angle n-HF.

54. CENTER OF A CIRCLE

To find the center of a circle passing through three given points using compass only, is none too easy. Two solutions are worthy of inclusion here. The first, by Theodore C. Andreopoulos, Structures Dept., Cornell Aeronautical Lab., Buffalo, New York, is quoted below: "To find the center of a circle passing thru 3 given points, using compass only, one proceeds as in the conventional solution employing straight edge. Thus, in Fig. 140, given points X, Y and Z, points A and B, on the bisector of XY, and points C and D on the bisector of YZ are found by using end points of the particular line and swinging any convenient radius. It now remains to determine the intersection of AB and CD. Construct symmetrics of C and D, that is C′ and D′, with respect to AB by using points A and B as centers and AC, BC, AD and BD as radii. Find E such that CC′BD is a parallelogram. Since DD′ and DE are both parallel to CC′, points D′ and E are collinear. Determine fourth proportional of D′E, D′D, and C′E (see Fig. 141). Using this as a radius and D′ and D′ as centers, find F. Since D′E/D′D = C′E/FD, and since D′F = FD, because AB is by construction the bisector of D′D, it is seen that F is the required center. It now remains to show

how the fourth proportional may be found by compass alone. In Fig. 141, draw concentric circles with radii D′E and D′D;

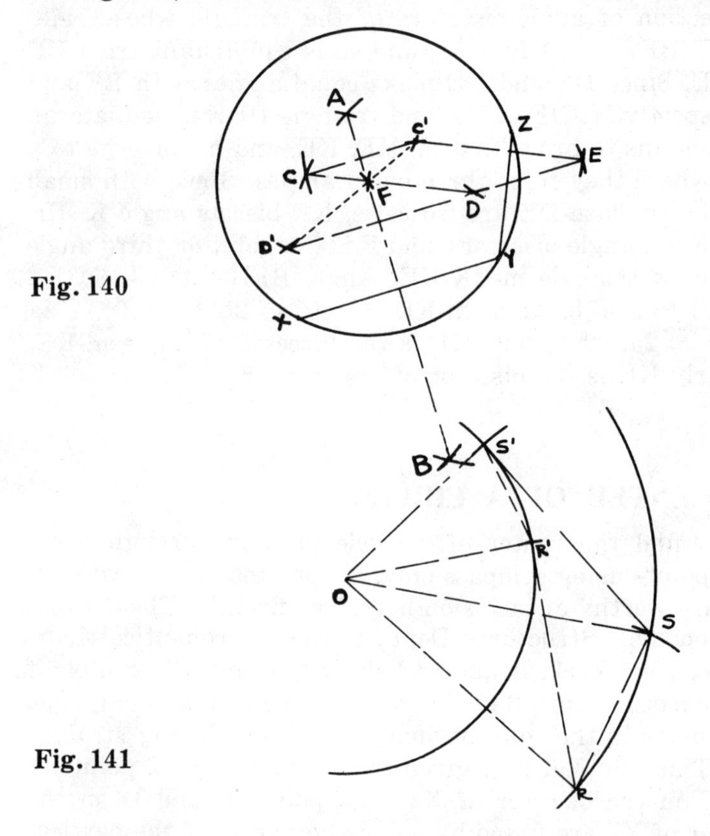

Fig. 140

Fig. 141

with C′E as a radius and any point R on the first circle as a center locate S on same circle. With any convenient radius and R and S as centers draw arcs intersecting the second circle at R′ and S′. Length R′S′ is the required fourth proportional. It is seen that triangles ORR′ and OSS′ are congruent. Therefore triangles ORS and OR′S′ are similar, and R′S′ is consequently the required fourth proportional". It is interesting to note that Mascheroni, the Italian geometer, demonstrated that all constructions possible by means of straight edge and compass can be accomplished by means of compass only.

The other solution, by F. G. Wilde, Paragon-Revolute Corp. makes use of inversion in neat fashion. Inversion is that branch of mathematics used in the study of a geometric figure which transforms it into its inverse figure, whereby certain propositions and properties, known as inverse propositions and properties, true for the original figure, yield propositions true for the inverse figure.

For example: The inversion of a point with respect to a circle, consists in finding the point on the radius drawn through the given point such that the product of the distances of the two points from the center of the circle is equal to the square of the radius — either of the points then being called the inverse of the other, and the center of the circle the center of inversion. Use of this principle actually shortens the construction in many cases, such as this. The solution by inversion follows:

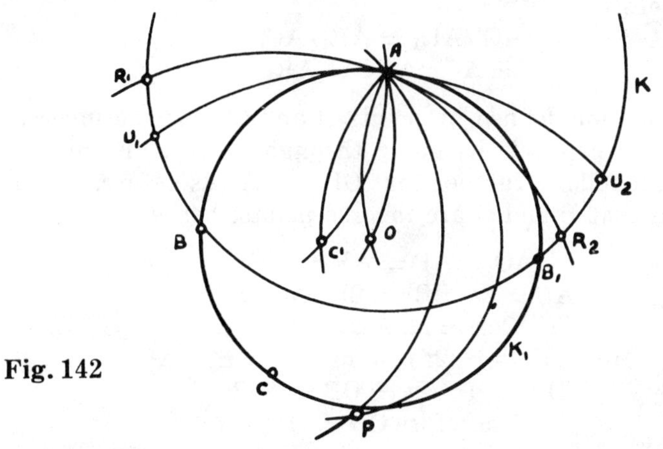

Fig. 142

With AB as radius and A as center, we draw a circle K (Fig. 142). With CA as radius and C as center, we describe an arc intersecting K at the points R_1 and R_2. With these two points as centers, we describe arcs with radius R_1A which intersect at A and C′. With centers B and C′ and radii BA and C′A respectively, we draw two arcs intersecting at P. With PA as radius and P as center, we describe an arc intersecting K at U_1 and U_2. With U_1 and U_2 as centers and U_1A as radius, we describe arcs intersecting at A and O. O being the center of the required circle K_1. The proof of this solution is based on the theorem of transformation through inversion, the analytical expression of which has the form $r_1r_2 = r^2$, where r_1 and r_2 are the distances of inverse points from a given point A, the center of inversion. A further condition of the inversion states that all inverse points have to lie on a straight line going through the center of inversion. It can be shown that after an inversion a line through A becomes a line through A, a line not through A becomes a circle through A, a circle through A becomes a line not through A, a circle not through A becomes a circle not through A, while points on the circumference of the basic circle of inversion are their own inverses. Selecting point A as center of inversion and selecting K as basic circle of inversion, it follows from the diagram that C and C′ must be inverse points. For triangles $AC'R_1$ and AR_1C are similar and therefore

$$AC'/AR_1 = AR_1/AC,$$
$$\text{i.e. } AC' \times AC = AR_1^2$$

Thus the line defined by points B and C′ must be inverse to the required circle K_1 going through A, B, C. From the diagram, it follows further that OB = OA. As before, it can be shown that P and O are inverse points, hence

$$AO \times AP = AV_1^2 = r^2$$
$$AP = 2a,\ BB_1 = 2m,\ AO = X$$
$$2ax = r^2,\ X = r^2/2a$$
$$OB^2 = (a - r^2/2a)^2 + m^2,\ m^2 = r^2 - a^2,$$
$$OB^2 = (r^2/2a)^2,\ OB = r^2/2a$$
$$\text{Therefore } OB = OA$$

Hence circle with center O and radius OA must go through B and therefore must be inverse to line BC′ and therefore it must also go through C and thus be the required circle K_1.

55. HOUSE ON A TRIANGLE

Although our DIAL readers ran true to form and submitted an amazing variety of widely different solutions, no one actually gave the simplest way to find the required length, which is to draw a triangle with sides a, b and c, construct an equilateral triangle on any one of the sides and measure the other diagonal of the quadrilateral so formed

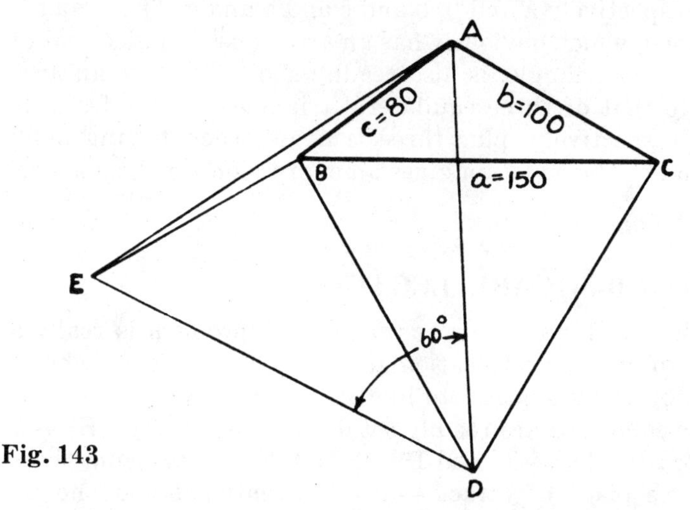

Fig. 143

(see Fig. 143). Proof that this is the desired length is had by merely rotating triangle ADC 60° about point D so that A goes to E and C to B. B is then a point distant respectively a, b, and c yards from the vertices of equilateral triangle AED, whose sides have the required length AD. Most readers started the other way around, with point B and unknown triangle AED and used the rotating technique in their solution. Incidentally, this rotation artifice has been used in a previous problem, the Pool Table Triangle Problem, No. 14. The algebraic solution is readily had from the above by applying the familiar cosine formula to give $d^2 = 1/2 (a^2 + b^2 + c^2 + \sqrt{3S(S-a)(S-b)(S-c)})$ from which d equals 179.6 yds. In so doing, one of our readers discovered a peculiar symmetry to exist in the relation between the side of an equilateral triangle and the distance of its vertices from a given point; namely $a^4 + b^4 + c^4 + d^4 = a^2b^2 + a^2c^2 + a^2d^2 + b^2c^2 + b^2d^2 + c^2d^2$, wherein any of the four quantities may be the triangle side and the other three the distances.

The result may also be had without the use of rotation or trigonometry by constructing on the three sides of the required equilateral triangle three triangles with their two sides respectively a and b, b and c and a and c. The hexagon so formed, which obviously has an area equal to twice that of the original triangle, is also readily shown to have an area equal to that of three equilateral triangles with sides a, b, and c respectively, plus three triangles each having sides a, b, and c. The equation gives an expression for d, as above.

56. FOUR-SQUARE LEGACY

The short-cut to the answer to this problem is really a source of much gratification to the mathematician who is clever, or lucky, enough to find it.

Four equations are readily set down: $J^2 - 2 = H$; $H^2 - 2 = -B$; $B^2 - 2 = -T$; and $T^2 - 2 = J$. From this point there are three plans of attack — two conventional and one inspired. 1) By successive substitutions an equation for a single one of the unknowns is readily obtained but this equation is of the sixteenth power, and the job of finding one or more correct numerical answers out of the possible 16 by successive approximations or graphical plotting is not simple, to say the least. 2) The four equations may be analyzed, as is, for short cuts in trial and error. It is evident, for instance, that John's and Tom's areas must both exceed 2 sq. miles, while both Henry's and Bill's areas must be less than 2. Starting with maximum and minimum lengths of 1.42 and 0 respectively for say Henry's plot, the corresponding maximum and minimum lengths are readily arrived at for Tom's plot between the narrow limits of 1.97 and 1.85. It takes only a few repetitions of the cycle to narrow down the value of T to the same figure for maximum and minimum to two decimal points, and it is readily shown that there is only a single solution where all the lengths are positive. 3) The inspired attack was employed by David G. Loomis of Evarts G. Loomis Co., Newark, N. J. and makes use of the fact that there is a subtle symmetry in the equations (we hinted at this in presenting the problem) in that each left hand member is the square of one of the unknowns, minus two, whereas the right hand member is one of the unknowns (plus or minus) to the first power. It happens that the square of the cosine of an angle bears a similar relation to the first power of the cosine of twice the angle, whereby $2 \cos^2 a - 1 = \cos 2a$. We therefore let $J = 2 \cos x$, from which our first equation becomes $2 (2 \cos^2 x - 1) = H$ and $H = 2 \cos 2x$. Substituting this value for H in the second equation and proceeding similarly for the next two develops the equation $\cos 16x = \cos x$ of which the general solution is $x = m\, 360°/15$ and $x = n\, 360°/17$, where m and n may be any integer from 1 to 8 to give a total of 16 different possible answers, conforming to our sixteenth degree equation. However, the conditions of the problem are such that cos x and cos 2x must be positive, and cos 4x and cos 8x must be negative, which requirements are satisfied only if $m = 1$, which means that $x = 24°$, and the answers are $J = 2 \cos 24° = 1.827$ miles; $H = 2 \cos 48° = 1.338$ miles; $B = -2 \cos 96° = .209$ miles and $T = -2 \cos 192° = 1.956$ miles.

Truly, it would be hard to find a more striking example of how analysis by the method of symmetry (see similar handling of problem No. 52, "The Inside Triangle") can shorten mathematical labors. In fact, these two examples might inspire the generality: When a math problem looks especially involved, see if things cannot be simplified by finding possible symmetry.

57. THE CONVEX LENS

The answer is readily arrived at by rearranging the equation in any one of a number of ways. Probably the simplest is to let $u = f + r$ and $v = f + s$, which is equivalent to measuring the object and image distance from the focal points instead of from the lens. The equation then becomes $rs = f^2$ and since r and s must obviously be integral in order

Fig. 144

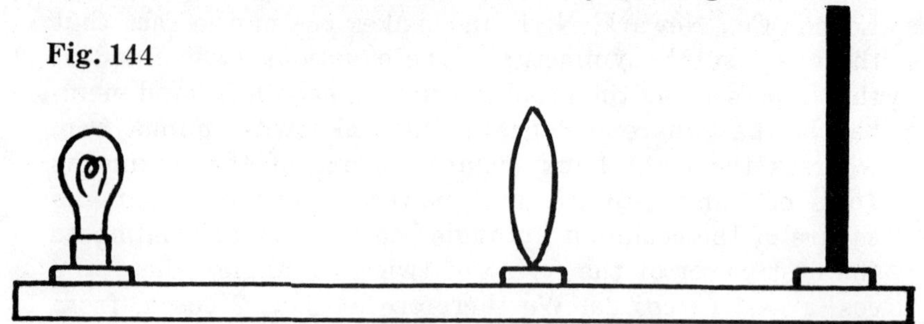

for u and v, which are respectively 12 greater, to be integral, Euclid needed only to find all the divisors of f^2 (in this case 144) and add 12 to each. The fifteen measurements are then 13, 14, 15, 16, 18, 20, 21, 24, 28, 30, 36, 48, 60, 84 and 156.

58. FIVES AND SEVENS

This is truly an intriguing illustration of the magic of numbers and strangely enough, the right mode of solution, which was used by Weston Meyer, Bundy Tubing Company, Birmingham, Mich. and by the author of the problem, harks back to that used in solving another intriguing puzzler in this book, No. 26. The Hatcheck Girl: "To find the chance that in a random distribution of hats by a hatcheck girl, no patron would get his right hat." It will be remembered that the first step in that solution was to discover that there was a definite algebraic relationship between the number of "satisfying" hat distributions X_n, X_{n-1} and X_{n-2} when the total number of hats was respectively n, $n - 1$ and $n - 2$ and then apply to the equation the known values for the simplest succession of cases, — namely $X_2 = 1$ when n is 2 and $X_1 = 0$ when n is 1. This led to a familiar periodic series for the required probability. Similarly, it may be shown in this problem by simple algebra that the successive roots t of the given equation $2X^2 - y^2 = 1$ always have the relation t_{n+2}

$= 6t_{n+7} - t_2$ for both X and y, whereby taking the first two pairs of roots $X_0 = 1$, $y_0 = 1$ and $X_1 = 5$, $y_1 = 7$, each succeeding root may be had by merely subtracting from six times the preceding root the root before that one, to give the values X = 1, 5, 29, 169, 985, 5741 etc. and y = 1, 7, 41, 239, 1393, 8119, etc. If we divide all X's by 10 and set down the remainders 1, 5, 9, 9, 5, 1 etc. and all y's by 7 and set down the remainders 1, 0, 6, 1, 0, 6 etc. we find on continuing the divisions that both sets of the six above remainders repeat themselves in periodic series, with a zero remainder for the division of the y root by 7 always corresponding to a 5 remainder for the division of the X root by 10, which was to be proved. The same method is used for 29 and 41 except that the first division is by 100 instead of 10. And the above is only a sample of the many strange relationships that can be culled from the Pell equation.

59. SIMILAR BUT NOT CONGRUENT

There are many solutions of varying complexity, but the most direct takes this simple approach: If the ratio be-

Fig. 145

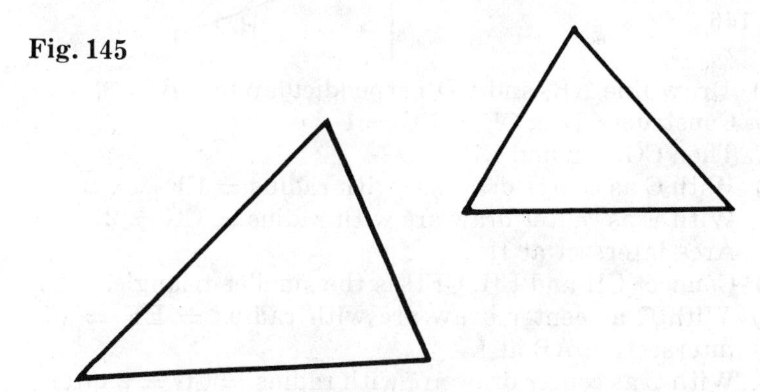

tween corresponding sides of the similar triangles is r, the sides can be set down respectively as a,b,c, and a/r, b/r, and c/r. Since the problem states that a = b/r and b = c/r (any other pairs of non-corresponding sides could have been selected as equal), we can rewrite a,b,c as b/r, b, and br, so that the requirement is met only if the lengths of the sides form a geometrical progression. The limiting cases are obviously when $r = 1$ and when r is such that $b/r + b = br$, since the sum of two sides of a triangle must be less than the third side. This reduces to $1/r + 1 = r$ from which $r = 1.618$. A typical solution, with $r = 1.5$ and with sides of integral length, is 8, 12, 18 and 12, 18, 27.

An interesting geometrical solution stated by E. F. Michel, G.E., Coshocton, Ohio, which illustrates the special case where $r = \sqrt{2}$, was probably the way little Euclid did the job for his dad and is illustrated in Fig. 146.

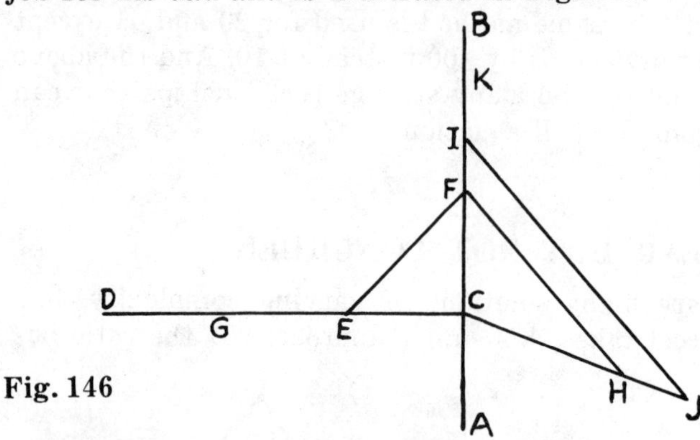

Fig. 146

1) Draw line AB, and CD perpendicular to AB
2) Construct $CE = CF = EG = 1$
   Then $CG = 2$ and $EF = \sqrt{2}$
3) With C as center draw arc with radius $= EF = \sqrt{2}$
   With F as center draw arc with radius $= CG = 2$
   Arcs intersect at H
4) Connect CH and FH. CFH is the smaller triangle.
5) With C as center draw arc with radius $= EF = \sqrt{2}$ intersecting AB at I.
   With C as center draw arc with radius $= CG = 2$ intersecting CH at J.
6) Connect IJ. CIJ is the larger triangle.
   To check the length of IJ, construct $CK = CG = 2$.
   $GK = IJ = 2\sqrt{2}$

60. MATCHSTICK SQUARES

This problem is essentially one of cut-and-try, the aim being, by Diophantine methods, to reduce the cut-and-try to a minimum, while achieving a rigid solution. Perhaps the simplest approach is to take advantage of the familiar relation that the sides of any integral right triangle, where the sides have no integral common factor, can be expressed as

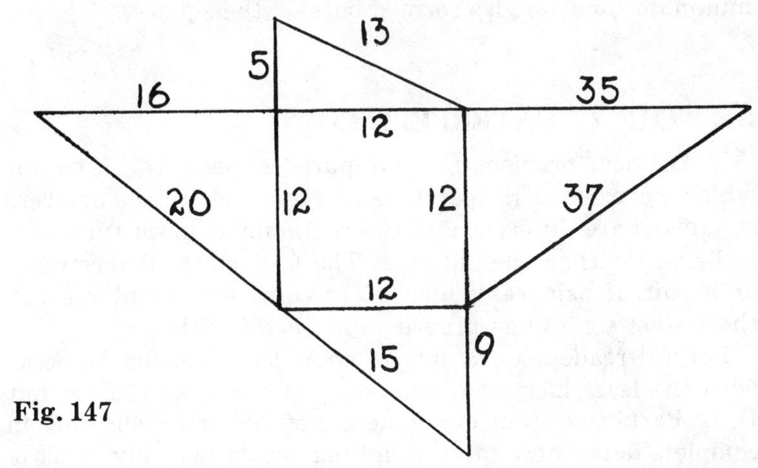

Fig. 147

xy, $(y^2 - x^2)/2$ and $(y^2 + x^2)/2$, where x and y are different positive odd integers. Starting with the smallest values of x and y namely (1,3) and proceeding upward in smallest increments to (1,5), (1,7), (3,5), (3,7) and (5,7) we get right triangles whose sides are respectively 3,4,5; 5,12,13; 7,24,25; 15,8,17; 21,20,29 and 35,12,37. Clearly, the length of side of the square we are looking for must, by the conditions of the problem, be divisible by four separate figures appearing in such tabulation. The smallest number meeting this condition is 12, which is divisible by both 3 and 4 in the first triangle, 12 in the second, and 12 again in the last; the corresponding triangles being 12,16,20; 9,12,15; 5,12,13 and 35,12,37, whose perimeters add up to 198. The same condition is satisfied by 15 and 20, which though larger than 12 could give a smaller combined perimeter, but the perimeters are found to total 226 and 298. For a check to see that no larger figure than these need be investigated we note that the perimeter $(y^2 + xy)$ of any right triangle in the above analysis exceeds twice the hypotenuse by $x(y - x)$ which excess is at least 2, and so must exceed twice any side by at least 2. If d is the required side of the square, this means that the sum of the four perimeters must be equal to or more than $4(2d + 2)$. If this total is to be 198 or less, d must be 23.75 or less, so we have exhausted all possibilities; and 198 is minimum for triangles formed outside the square.

61. THE COUNTERFEIT COIN

An ideal problem for our purposes, and the kind for which we are ever in search, is one in which all the answers submitted are different and the optimum solution turns out to be better than the author's. The problem of determining on a pair of balances which of 12 coins was counterfeit in the fewest weighings proved to be in this category.

Several readers were able to show that it could be done, with the least luck, in a maximum of three weighings, but D. B. Parkinson went everyone a step better by showing in complete detail how three weighings could not only disclose the counterfeit, but also whether it was over or underweight. This solution was also used by Lester H. Green, of Detroit, whose ingenious, graphic handling portrays Mr. Parkinson's method of solution in unique fashion (see Fig. 148).

The figure is largely self-explanatory. The conclusion reached after each weighing is given below the lines leading to the next weighing, for each possible outcome. Briefly, in the first weighing, two sets of four coins are balanced against each other and the other four not weighed. If the two sets balance, the bad one must be among the other four, three of which are then weighed against three good coins, and only one additional weighing is then needed to tell which of the four is bad, and whether it is light or heavy. If the two sets don't balance, the trick is then to take three from the right side, plus one from the heavy side, and weigh them against the fourth coin from the light side plus three good

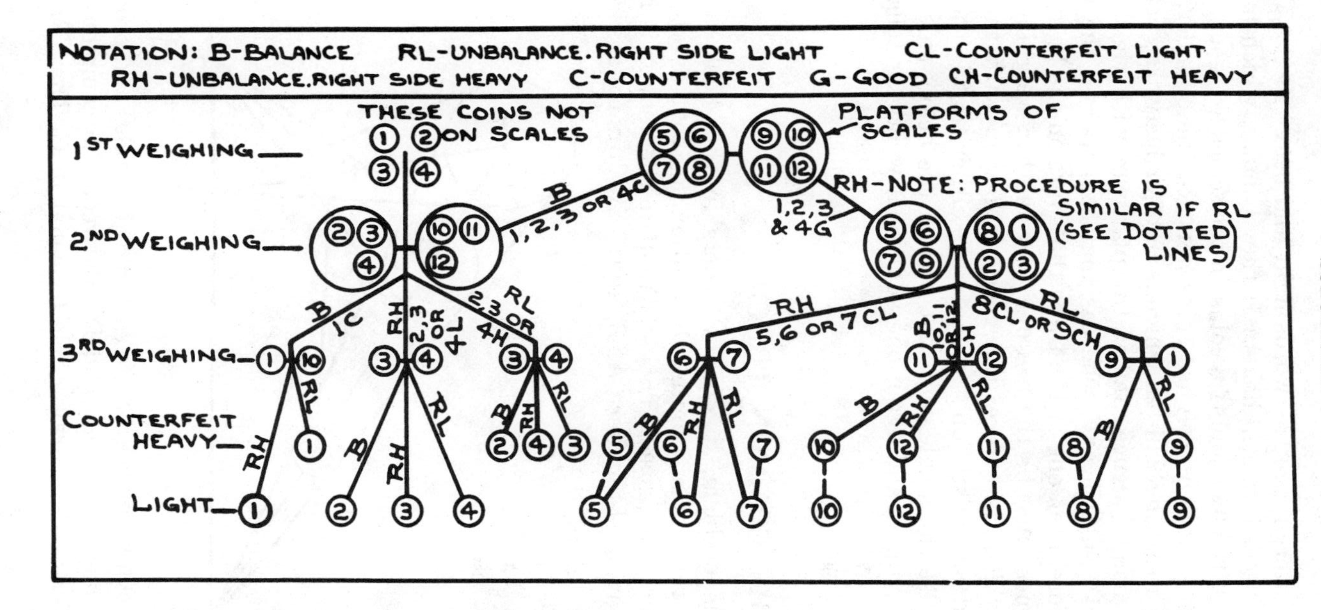
NOTATION: B-BALANCE RL-UNBALANCE. RIGHT SIDE LIGHT CL-COUNTERFEIT LIGHT
RH-UNBALANCE. RIGHT SIDE HEAVY C-COUNTERFEIT G-GOOD CH-COUNTERFEIT HEAVY
THESE COINS NOT ON SCALES
PLATFORMS OF SCALES
1ST WEIGHING
2ND WEIGHING
3RD WEIGHING
COUNTERFEIT HEAVY
LIGHT
B 1,2,3 OR 4C
1,2,3 & 4G
RH-NOTE: PROCEDURE IS SIMILAR IF RL (SEE DOTTED LINES)
B 1C
RH 2,3 OR 4L
RL 2,3 OR 4H
RH 5,6 OR 7CL
B 10,11 OR 12CH
RL 8CL OR 9CH

Fig. 148

coins. (Interchange the words light and heavy in the above and the procedure still holds.) If they balance, which means that all of the five doubtful coins on the scales are good, then one of the three unweighed coins must be the offender, and another weighing will tell which is light or heavy, as before. If they don't balance, and if the platform with the single doubtful coin is heavy, whereas that coin was on the light side before, then one of the three that were on the light side both times must be the bad one, and another weighing discloses which, as before. However, if the pan with the one doubtful coin is still on the light side, either that is the bad one or else the single coin that was on the heavy side both times is guilty, and again one additional weighing tells the story.

62. THE COMMON TOUCH

This problem, like many others in this book, allows various roundabout routes to solution, in contrast to the single direct approach. This short-cut to the answer is illustrated in Fig. 149 where A, B, and C are 3 unequal circles.

Fig. 149

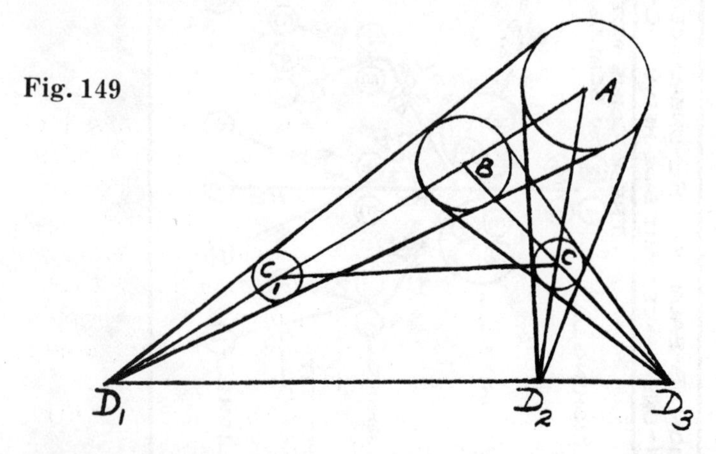

Draw common tangents to each pair of circles and produce them to meet at D_1, D_2 and D_3. Between the common tangents of A and B, draw circle C_1 equal to C. Then Radius

A/Radius C $=$ AD_1/C_1D_1 $=$ AD_2/CD_2 or AD_1/AD_2 $=$ C_1D_1/CD_2. Therefore C_1C is parallel to D_1D_2. Similarly, C_1C is parallel to D_1D_3. Therefore D_1D_2 and D_1D_3 are coincident and are one straight line.

It is related by K. K. Knaell of the Federal Glass Co., Columbus, Ohio, who furnished the above solution, that when this problem was originally presented to Prof. Sweet he paused a moment and said, "Yes, that is perfectly self evident." Astonished, his friend asked him to explain how a problem that stopped most advanced students could be called self evident. Prof. Sweet, in effect, replied, "Instead of three circles in a plane imagine three balls lying on a surface plate. Instead of drawing tangents, imagine a cone wrapped around each pair of balls. The apexes of the three cones will then lie on the surface plate. On the top of the balls lay another surface plate. It will rest on the three balls and will be necessarily tangent to each of the three cones, and will contain the apexes of the three cones. Thus the apexes of the three cones will lie in both of the surface plates, hence they must lie in the intersection of the two plates, which is of course a straight line."

63. HIGHEST APEX

Queerly enough, this problem, which seems at first like a simple exercise in analytical geometry (but if so handled can follow needlessly circuitous routes), harks back to our No. 38, "Captivating Problem in Navigation". This told of a ship making 15 knots being pursued by a ship going twice as fast, the pursued ship taking advantage of a cloud bank to change its course abruptly. The problem was to determine the proper course of the pursuing captain (who guessed rightly that the pursued would stick to his changed course), to insure capture. Here, too, the locus of the point where the clever captain would start spiralling logarithmically was an "Apolonius" circle which defines a point whose distance from a fixed point (where the pursued entered the cloud bank) bears a given ratio, here 2 to 1.

When tackled as a conventional exercise in analytical geometry one arrives in due course at the circle — as in our navigation problem — which for the figures given had a radius of 12″, which is the maximum height, and for the general case $k/(k^2 - 1)$.

By this approach, for the special case given we quote from a typical solution of this kind. The base is placed on the x axis so that one end is at the origin, and the other end is at point 10, 0. Then we let point x, y be any point whose distances from the ends of the base are in the ratio of 3/2. According to the distance formula we find that the distance from point x, y to point 0, 0 (the origin) is $\sqrt{x^2 + y^2}$ and the distance from point x, y to point 10, 0 is $\sqrt{(x - 10)^2 + y^2}$. Setting these distances equal according to the given ratio, we have the equation

$$3\sqrt{x^2 + y^2} = 2\sqrt{(x - 10)^2 + y^2}$$

which is the equation for the line on which the apex must lie to satisfy the given conditions. Squaring both sides and simplifying, we get $(x + 8)^2 + y^2 = 144$, which is the equation of a circle whose center is at point −8, 0 and whose radius is 12.

To find the equation for the general case, we use the same procedure, but substitute a and b for 3 and 2, and use 1 (unity) for the base, instead of 10. This yields the following equation:

$$a\sqrt{x^2 + y^2} = b\sqrt{(x - 1)^2 + y^2}$$

Squaring and simplifying, we again have the equation of a circle:

$$\left(x + \frac{b^2}{a^2 - b^2}\right)^2 + y^2 = \frac{b^2 a^2}{(a^2 - b^2)^2}$$

from which the center is at $-\frac{b^2}{a^2 - b^2}$, and the radius is $\frac{ab}{a^2 - b^2}$.

Some readers of the DIAL, after expressing y in terms of x, failed to notice that this relation represented a circle and set the derivative dy/dx equal to zero to get the maximum value. How far afield one can go by "doing what comes naturally!"

However, instead of using any equations in analytical geometry, a neater approach is to take the given line CD

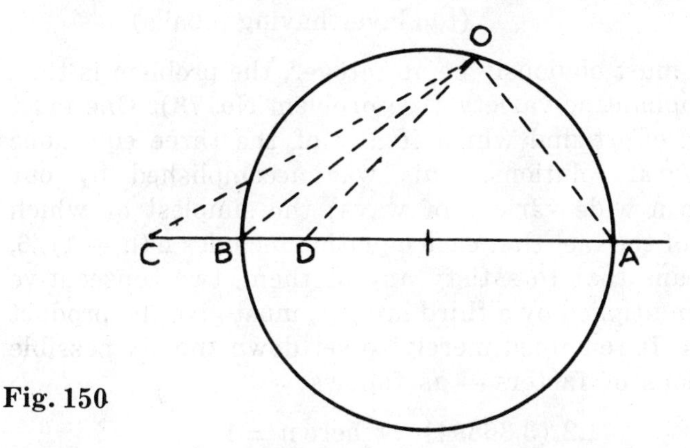

Fig. 150

(Fig. 150), assume any point O where OC/OD equals k and draw OB and OA, the bisectors of the internal and external angles COD. By familiar geometry, both CB/CD and CA/DA equal k, so that B and A are fixed points. Since angle BOA is obviously a right angle, the required locus is a circle whose diameter is BA. Taking $CD = 1$, $DA = (1 + DA)/k$ from which $DA = 1/(k - 1)$ and since $BD = 1/(k + 1)$, the sum of the two, which is the required diameter, is $2k/(k^2 - 1)$ as before.

64. PYRAMID OF BALLS

When this problem appeared in the DIAL dozens of readers arrived at the correct answer but most of them (including, we must admit, the author of the problem) found it necessary to go into rather extensive algebra and cut-and-try to reach the result. Clearly, the first step is to set down (by familiar algebra for the sum of arithmetical progressions of the second order) formulas for the total number of balls in each type of pyramid, in terms of the number of layers n, as follows:

$$\text{Triangular: } S = n(n+1)(n+2)/6$$
$$\text{Square: } S = n(n+1)(2n+1)/6$$
$$\text{Rectangular: } S = n(n+1)(2n+3p-2)/6$$
$$\text{(top layer having p balls)}$$

Since n must obviously be an integer, the problem is then of the Diophantine variety (see problem No. 73). One must with least effort find which if any of the three equations have integral solutions. This was accomplished by our readers in a wide variety of ways; the simplest of which made use of the fact that each equation includes $n(n+1)/6$, which means that to satisfy any of them, two consecutive integers, multiplied by a third integer, must give the product 6 x 36,894. It remained merely to set down the six possible combinations of factors — as follows:

$$1.2.(3.36894) \quad \text{where } n = 1$$
$$2.3(36894) \quad \text{where } n = 2$$
$$3.4(18447) \quad \text{where } n = 3$$
$$11.12.(3.13.43) \quad \text{where } n = 11$$
$$12.13.(3.11.43) \quad \text{where } n = 12$$
$$43.44.(3.3.13) \quad \text{where } n = 43$$

It takes only a glance at the factor in the parentheses to see that it does not in any case equal either $(n + 2)$ or $(2n + 1)$, which automatically throws out the possibility of either a triangular or a square pyramid. This leaves for consideration only a rectangular pyramid. When $n = 2, 3, 11$ or 12, p is fractional, which eliminates all possibilities except $n = 1$ and $n = 43$. When $n = 43$, $p = 11$ which gives a rectangular pyramid of 43 layers and 11 balls on top. When $n = 1$, p is the integer 36,894 which gives the limiting case of a rectangular pyramid of one layer and 36,894 balls in a line.

The method above was followed by many readers one of whom made use of the interesting fact that since 6 x 36,894 is not divisible by 10, either n or $n + 1$ would have to be prime. Another solution which arrived by a somewhat longer route at the same rectangular pyramid having a base of 43 balls by 53 balls, stated that the pyramid "will be 30.7″ high and weigh about 5460 pounds, putting a load of 360#/sq. ft. of floor. "That floor," he warned, "had better be adequately braced."

65. ALICE AND THE MAD ADDER

This presents a Diophantine task of finding four factors of 711,000,000 which add up to 711, the job being simplified by expressing the amounts in cents. In order to give the six zeros in the final product, the result of any individual multiplication must end in zero. Since the other two figures could not furnish more than 5 zeros, the first figure of such product then not being 2 or 5, and since the unit digits must add up to 1; the only possibilities are 0, 0, 0, 1 and 0, 0, 5, 6. The factors of 711,000, are $(2)^6$ $(3)^2$ $(5)^6$ (79), and in order to fulfill the above conditions 79 must be multiplied by 4, since if multiplied by 5 the four factors would have to add up to more than 711, because even if the other three factors could then be equal, which gives the smallest sum, they would add up to more than 711 minus 395. This derives from the general relation that the factors of a number add up to the least amount when they are equal. We therefore know that one factor is 316 and the other three factors which end in 5, 0, and 0 respectively add up to 395. Since the cube root of the remaining product is 131, none of the three factors can depart much from this average. This leaves only a few combinations to try before arriving at the figures of 1.25, 1.20, and 1.50.

66. THE WALK AROUND

This problem is the kind that gives special satisfaction because it ostensibly requires a complicated mathematical solution, but actually can be solved by simple reasoning if one pursues the right approach. Here again, the path was such as Bunyan might have dubbed "the easy road of symmetry." Conventional mathematics would tackle the problem

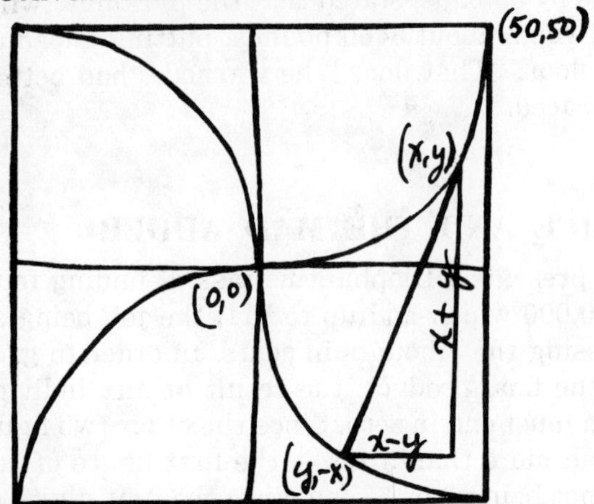

Fig. 151

as seen in Fig. 151, in which the curve in the upper right quadrant is defined by $dy/dx = (x + y)/(x - y)$. Transforming to polar coordinates by $x = r \cos a$, $y = r \sin a$, $dy = \sin a\, dr + r \cos a\, da$, $dx = \cos a\, dr - r \sin a\, da$, we get the solution $a = \log r + c$ or $r = c^{1} e^{a}$ (logarithmic, not Archimedean spiral) where the constant need not be determined. The arc length of this curve from the origin to (50, 50) is found by integrating between the limits of $r = 0$ and $r = 50\sqrt{2}$ the expression $\sqrt{dr^2 + r^2\, da^2}$ which gives the answer 100, whereby Euclid would get back to the center at the same instant as his four chums.

However, it is possible to get the identical result, with equal rigidity but less labor, by taking the aforesaid road of symmetry as follows: Arthur, Ben, Charles, and David initially form a square and because of symmetry must always be in positions to form a square until they meet in the center (Fig. 152). Hence, initially and always, the motion of each is perpendicular to the instantaneous path of his pursuer. In effect, therefore, the motion of the pursued does not alter the distance between him and the pursuer. (In other words, it is as if each was walking toward a stationary object, and the rotation and shrinkage of the square defining the boys' positions have no bearing on the distance each must travel.)

Consequently, the pursuer reaches the pursued only after traveling the whole intervening distance that prevailed at

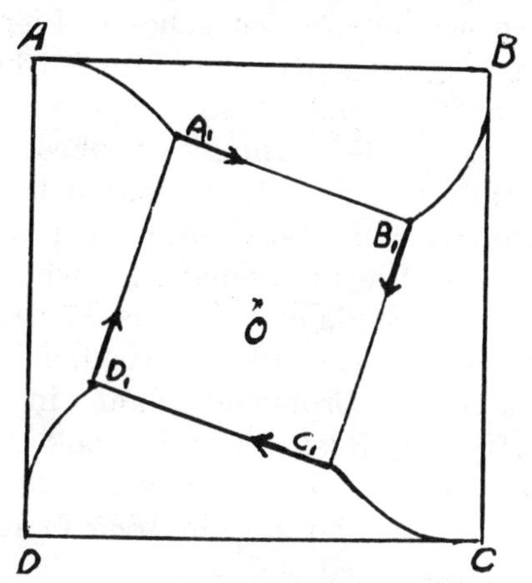

Fig. 152

the start. The distance each travels, then, is 100 feet, the side of the square.

67. THE ABSCONDER'S LEGACY

Mr. Lincoln's fascinating problem of the "Absconder's Legacy" proved a stumbler for many of our readers. Expressed briefly, a fund was accumulated by setting aside one cent the first day, two cents the second day, three cents the third day, and so on until it attracted an absconder who found upon counting his haul later in the Tropics, that it was just large enough to buy each of his twenty-three grandchildren a quantity of baboons equal to the price per baboon in cents — the question being how many days had the fund accumulated before it was lifted.

If the number of days is n and the price per baboon B, then by the familiar relation for the sum of an arithmetical progression, $n(n + 1)/2$ must equal $23B^2$ and by setting x as equal to $2n + 1$ and y equal to $2B$, we get the equation in the more usable form $x^2 - 46y^2 = 1$. This indeterminate equation with two integral unknowns is of the kind that Diophantus wrestled with so valiantly back in the third century. It also happens that in the general form $x^2 - Ay^2 = 1$, it made mathematical history a mere fourteen centuries later when a profound French scholar, Pierre de Fermat, challenged two English mathematicians in the year 1657, to find the general solution. (It so happened that a century later the renowned Euler wrongly ascribed this equation to John Pell instead of to Fermat so that it became known as the Pell equation. As a matter of interest it may be remarked that John Pell was the mathematician who introduced the division sign $\div$ in England. He was Professor of Mathematics at Amsterdam from 1643 to 1646, and at Breda from 1646 to 1650, and was Cromwell's agent in Switzerland for some years. After the Restoration he took orders and had a living in Essex. He was particularly versed in Diophantine analysis. Sad to relate, he died in 1685 in London in great poverty.) Fermat showed that the equation $x^2 - Ay^2 = 1$ has an infinite number of solutions, which fact is equivalent to the surprising truth (one of the most fascinating in all number lore) that any non-square can be multiplied by any one of an infinite number of squares, so as to give a product one less than another square. For instance the non-square 3 when multiplied by the square 1 gives the product 3 which is one less than the square 4; also, 3 when multiplied by the square 16 gives the product 48 which is one less than the square 49; when multiplied by the square 3136 it gives 9408 which is one less than the square 9409, and there are an infinite number of such cases, although the required squares soon become astronomical in size.

To come back to our equation $x^2 - 46y^2 = 1$, it is obvious that the ratio x/y is slightly more than the square root of 46. The exact value is found by setting up the root of 46 as the continued fraction (familiar to students of higher algebra):

$$6 + \frac{1}{+1}\,\frac{1}{+3}\,\frac{1}{+1}\,\frac{1}{+1}\,\frac{1}{+2}\,\frac{1}{+6}\,\frac{1}{+2}\,\frac{1}{+1}\,\frac{1}{+1}\,\frac{1}{+3}\,\frac{1}{+1}$$

and trying each convergent until the equation is solved. In this case the twelfth convergent shows x to be 24335 and y to be 3588 so that the required number of days n is 12167 — or almost exactly one-third of a century. This answer was arrived at in the above manner by Francis L. Miksa of Aurora, Illinois, and Howard D. Grossman of New York, both of whom showed also that the next solution, in which y is readily set down as the product of 24335 × 3588 × 2 or 174,627,960 gives an answer of 592,192,224 days which is well over a million years and is thus ruled out. Incidentally, various methods of less scientific cut-and-try were used by other readers to get the correct answer.

Of these solutions the most ingenious was by Otto F. Schaper, Port Washington, N. Y. who wrote "My first approach to the 'Problem of the Absconder's Legacy' was the following. The basic equation is $\frac{x(x+1)}{2} = 23y^2$ or $x(x+1) = 46y^2$. This is indefinite because there is no mathematical equation to cover the condition of integer numbers. I therefore substituted $y = \frac{u}{v}$.

$$x(x+1) = 46y^2 = \frac{46}{v}\frac{u^2}{v} \text{ so that } \frac{u^2}{v} - \frac{46}{v} = 1 \text{ would}$$

be the answer provided that $\frac{46}{v}$ is an integer, in which case $u^2 - 46 = v$. x is expected to be a large number which calls for v to be small. I set my slide rule to root of 46 and looked for fractions very nearly equal to that root. $\frac{156}{23}$ got my interest because the basic equation has prime number 23 as a factor and sure enough, with that value for u, v is $\frac{2}{529}$, y is 1824 and x is the required integer 12167. Here the approach by 'reasoning and feel' rendered the solution within 20 minutes and it is frequently successful in indefinite equations and mathematical puzzles. But if the first trials fail it is usually not suitable for further steps."

68. PILOTS' MEETING

In presenting this problem we departed from the original version as contributed by H. S. Pardee, in which the officers after taking off at 2 P.M. were all to land at the same time and place. We wanted to avoid the involved trigonometry and felt that some practical-minded readers might question the delayed meeting. Nevertheless, several who did not read the problem too carefully actually solved the more com-

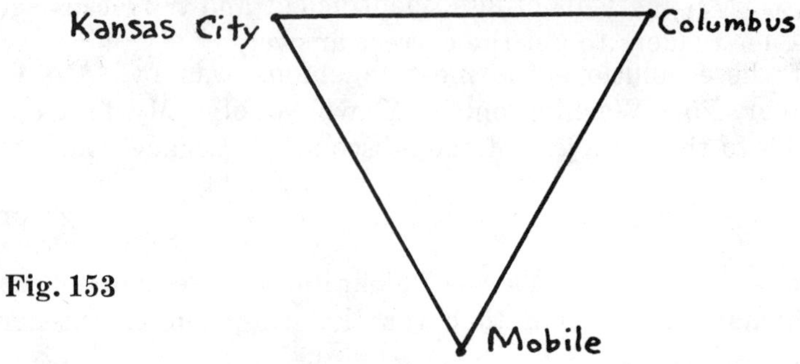

Fig. 153

plicated version, and we include one such solution (see Fig. 154) as a matter of interest. Here, for simplicity, the distances are taken as unity and the time t in hrs/100. Applying the law of cosines to triangles CAP and BAP, P being

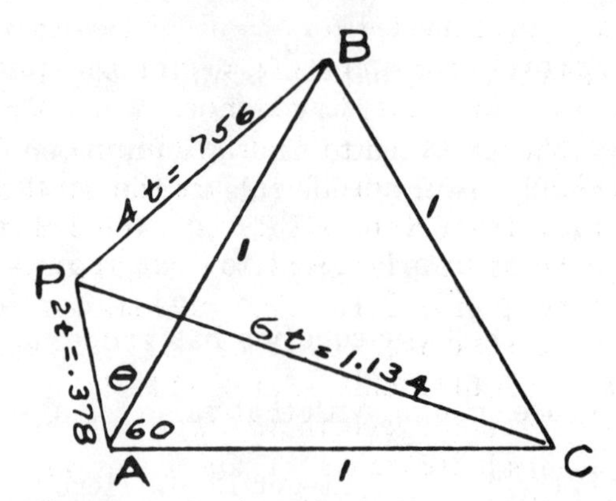

Fig. 154

the meeting point, we have $CP = 6t = \sqrt{(1 + 4t^2 - 4t \cos (60 + \theta))}$ and $BP = 4t = \sqrt{(1 + 4t^2 - 4t \cos \theta)}$ which gives $t = .189$ or, for the given distance of 700 miles, $t = 1.323$

hours or 79.38 minutes, θ being 40° 53.51′ and AP = 264.6 miles, which locates the meeting point.

The problem as stated is correctly solved as follows: "The meeting place most quickly reached would be on the straight line between the starting points of the two slowest planes, providing that the 600-mph jet could arrive simultaneously or before, since otherwise the two slow planes would have farther to go. As the jet travels at a speed equal to the sum

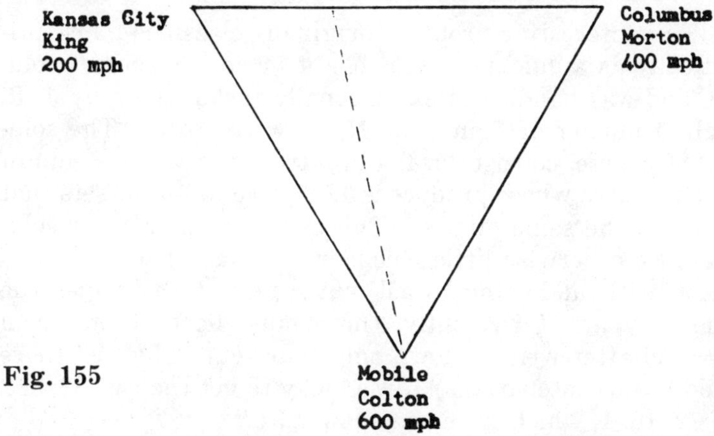

Fig. 155

of the speeds of the other two planes it could cover 700 miles in the shortest time needed for the other two to meet. Since their line of travel is the opposite side of an equilateral triangle from the point located at Mobile, no point on this line can be at a greater distance from Mobile than 700 miles, which is the length of a side. Therefore the point of meeting is 233-1/3 miles from Kansas City and 466-2/3 miles from Columbus. Both of the slower planes would arrive 1 hr. 10 min. after starting or at 3:10 P.M. The distance from Mobile to this point is a little over 617 miles, so Colton would arrive in approximately 1 hr. 2 min. or at 3:02 P.M."

If the speeds are changed to 300 mph, 400 mph, and 500 mph, the essential difference in the situation is that the speed of the fastest plane is then not great enough to permit its meeting the other two planes on the line joining their starting points. As a result, the earliest meeting point was within the triangle and all three planes would get there simultaneously because there is only one point within a triangle whose distances from the vertices are respectively proportional to three given amounts. At any point other than this, at least one of the planes would have further to go — delaying the meeting.

69. THE CENSUS TAKER

This interesting problem, originally classified a bit optimistically as a quickie, is capable of various modes of solution, and was handled in exceptionally fast fashion by J. R. Hugh Dempster of Princeton, N. J., who wrote: "The solution, of course, consists in finding, from the various sets of three integers whose product is 1296, two different sets that add up to the same number. This must be the correct solution since otherwise little Euclid would have had his information without having to ask, "Are any of their ages the same as yours?" Evidently, there must be only one such case, and after writing down only four sets (I figured there would be no centenarians) I was lucky to get the two 81, 8, 2 and 72, 18, 1, which gave the answer 91."

70. TRIANGLE OF COINS

As in the famous game of Nim (No. 30) the trick is to set up a balanced situation with the powers of 2 appearing an even number of times, such as — twice or not at all, rather than once or three times, and maintain this tactic till a favorable end situation is reached, said end situation depending on whether you or the partner must take the last coin to win. Initially the powers of two are as in the diagram, with the fours and twos paired but the ones unbalanced so that the first player should take a single coin from the first, third or fifth row. If he isn't the one in the know and takes say 2 coins from the middle row, a balance is then obviously reached by the knowing partner by taking one coin from the second row. This procedure is continued until the knowing player sets up an obviously winning end situation which depends on the objective and in some cases is unbalanced, such as leaving 1, 1, 1, or 1, 1, 1, 1, 1 if the taker of the last coin is to be the loser. It is important to note that if the man "in the know" draws

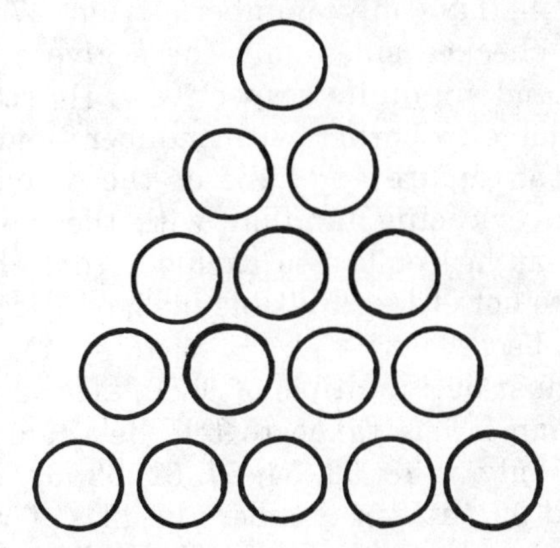

Fig. 156

second, after the first man has inadvertently put the rows in a safe condition, he will have to draw a single coin in the hope that in the second round the first man will err. Inci-

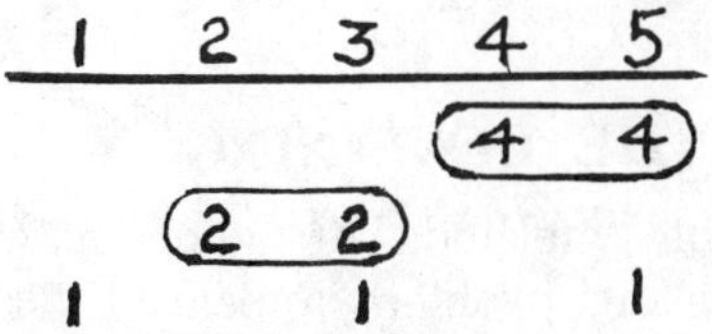

1	2	3	4	5
			4	4
	2	2		
1		1		1

Fig. 157

dentally, with as few coins as 15, the equivalent of the above could, of course be arrived at empirically by the tyro without reference to binaries, that is, ones twos and fours. The expert, however, could then confound him by changing the number of rows and/or coins per row to any random figures and follow the same procedure as above.

71. SQUARE PLUS CUBE

Obviously, the quickest way to get the answer is by looking up a table of squares and cubes in a handbook. But, with a few seconds' thought before so doing, the time to make the check — and recheck — may be reduced a bit, with certainly an increase in the enjoyment. At any rate, that was the procedure followed by George Karnofsky, of Blaw Knox, Pittsburgh, who, noticed — as did all who solved the problem — that only numbers from 47 to 99 inclusive need be checked since only these give a square and a cube of four and six digits respectively. He realized that he also did not have to bother with numbers ending in 0, 1, 5 and 6 since the square and cube of these end in the same figure. Moreover, being familiar with the law of nines, he recognized that he could also exclude from check all numbers that were not either divisible by 3, or 1 less than a multiple of nine, because $x^3 + x^2 = x^2 (x + 1)$, one of which two factors must be a multiple of 9. So, after a pause shorter (we hope) than it has taken to tell, he thumbed down the list, stopping only at 48, 53, 54, 57, 62, 63, 69, 72, 78, 84, 87, 89, 93, 98 and 99 (fifteen numbers in all), of which only 69 fills the bill. Incidentally, Prof. Thebault, who discovered this relationship, could not explain why there is just a single solution — but this makes the poser more unique and intriguing.

72. END AT THE BEGINNING

Three possible methods of solution were uncovered when this interesting problem appeared. The conventional approach by algebra sets up the relation such that if the original number is taken as $(10a + b)$ with a total of $(n + 1)$ digits, then $9(10a + b) = 10^n b + a$ from which $89a = (10^n - 9)b$. Since $10^n - 9$ is, of course, 9999 . . . 1, we divide this by 89 until there is no remainder when the 1 is brought down, which gives the quotient a. n turns out to be 43 and since b is obviously 9, the required number, which was likened by many readers to the national debt, is the astronomical figure 10,112,359,550,561,797,752,808,988,764, 044,943,820,224,719. Two other modes of solution avoid the long division by 89. One accomplishes the job by short divi-

sion. Since the first and last digits of the required number must be 1 and 9 respectively, the first two figures of the number when multiplied by 9 are 91. We can then set down the divisor $9)\underline{9101-}$, putting the previous figure in the divi-
101

sor in the dividend each time, and continuing until there is no remainder when the figure 1 is set up in the dividend. A reverse procedure, using multiplication by 9, takes into account that the last two figures of the original number must be 19 and sets up the multiplication

$$\begin{array}{r} 9 \\ -4719 \\ \hline 471 \end{array}$$

, putting the previous figure in the product in the multiplier each time until the figure 1 appears in the multiplier without carry-over.

73. POLITICS IN THE STATE OF CONFUSIA

This is obviously a problem of the Diophantine variety (see also Nos. 35 and 64). By choice of proper parameters, we can find in the simplest way, that is, with least cut and try, the integral values of an unknown quantity which satisfy an equation that is further limited by certain restricting conditions. Diophantus was the great algebraist of some sixteen centuries ago, about whom it has been written

Diophantus' fame will never die,
Parameters tell the reason why,
He showed how to use them to simplify
The arduous job of "cut-and-try."

In this case, the job is obviously to find a final number of states closest to 275, which in connection with a properly chosen plurality will give ten successive numbers of states that will in each case result in the same plurality. If the number of states is n, if the number of representatives voting for the winner is $n - x$ and those voting for the loser x then $(n - x)^2 - x^2 = P$ (the constant plurality). This may be written $n(n - 2x) = P$. There must be 10 different values

	Number of states	Winner	Loser	Plurality P
A	n	$n - x$	x	$n(n - 2x)$
	240	127	113	240 (14)
	210	113	97	210 (16)
	168	94	74	168 (20)
	140	82	58	140 (24)
	120	74	46	120 (28)
	112	71	41	112 (30)
	84	62	22	84 (40)
	80	61	19	80 (42)
	70	59	11	70 (48)
	60	58	2	60 (56)
B				
	270	143	127	270 (16)
	240	129	111	240 (18)
	216	118	98	216 (20)
	180	102	78	180 (24)
	144	87	57	144 (30)
	120	78	42	120 (36)
	108	74	34	108 (40)
	90	69	21	90 (48)
	80	67	13	80 (54)
	72	66	6	72 (60)
C				
	252	136	116	252 (20)
	210	117	93	210 (24)
	180	104	76	180 (28)
	168	99	69	168 (30)
	140	88	52	140 (36)
	126	83	43	126 (40)
	120	81	39	120 (42)
	90	73	17	90 (56)
	84	72	12	84 (60)
	72	71	1	72 (70)

of n, the biggest of which is very nearly 275, which satisfy this equation. Clearly, the second factor cannot be larger than the first since that would make x a negative quantity. Also, the factors are either both odd, or both even. There must be enough prime factors of P to admit 10 different combinations of the two factors, starting with n nearly 275 and finishing with n no less than 2 more than the other factor. To accomplish this, the maximum value of n should have as many small prime factors as possible to allow for the variation. The value of P as compared with this maximum, should be as small as possible, so that it introduces the minimum number of small prime factors beyond those of the maximum value of n. It is readily found that the three different tables of values of n, $n - x$, x and $n\ (n - 2x)$ that meet these conditions can be formed as below; the second table, with a top value of $n = 270$, is obviously the optimum solution.

74. FIDO-TABBY CHASE

Mr. Johnson's provocative problem of the Fido-Tabby chase, wherein Fido defies the teachings of Euclid and Pythagoras, not only brought some excellent answers but also evoked some profound philosophy. In offering this problem, we had remarked in passing that Fido, if possessed of the knowledge and foresight of the more articulate so-called human race, would have instantly set this course at N. $\sin^{-1}$ 4/5E., and saved himself a lot of running. This brought a letter of protest from Fido himself, writing from Montreal in the guise of Roland Bourjault of the Quebec Hydro Electric Commission, as follows: "Gentlemen: Who is trying to insult me, Fido, in saying I am not possessed of 'human knowledge and foresight?' At least, I do hope man does not think he can catch a cat more easily than I do. Sure, I know Pythagoras' theorem, but better than man, I put it in actual practice. Instead of shooting a straight line to a hypothetical point, I keep my distance from the cat at a minimum at all times, because (1) I know I might lose some speed, or gain some on the last stretch, and (2) (which is the most important point) I am not dumb enough to take a chance on the cat retaining her straight and constant due east course. What happens to my Pussy Meal if I use this 'human knowledge,' when Pussy decides some other course is better for her health?" To which we humbly replied: "Dear Fido: We have your letter of protestation. It was doggone good. But please note that we kept our own tongue in cheek when we wrote 'so-called' human race. At any rate, we hope you enjoyed your Pussy Meal — or melee, whichever it turned out to be."

To find the penalty paid by Fido in yardage run for keeping the prey constantly on his course, requires the aid of Newton or Leibnitz. Solving first in general terms, (see Fig. 158) we let D equal the distance between dog and cat at any instant, a the angle between the dog's pursuit curve and the cat's track, v the speed of the cat and cv the speed of the dog. Then the relative approach speed $dD/dt = -(cv - v\cos a)$ and the angular speed of the dog $da/dt = v \sin a/D$ from which $vdt = dD/(c - \cos a) = - Dda/\sin a = - (MdD - NDda)/M(c - \cos a) - N \sin a$ where M and N can of course be any arbitrary values. Setting for convenience $M = c + \cos a$ and $N = - \sin a$ gives $vdt = - (cdD - \cos a\, dD - \sin a\, Dda)/(c^2 - \cos^2 a - \sin^2 a)$ which re-

Fig. 158

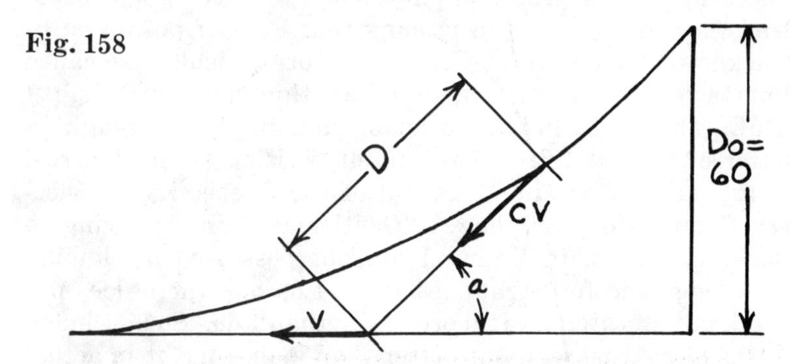

duces to $- cdD - d(D\cos a)/c^2 - 1$. Integrating between the limits $D = O$ and $D = Do$ gives the escape distance $cDo/c^2 - 1$, which yields the rather simple and striking relation that the distance traveled by the pursued equals the initial distance between pursuer and pursued times a fraction equal to the ratio between their speeds divided by a number one less than the square of that ratio. In this case the escape distance equals

$$\frac{\frac{5}{4}(60)}{\left(\frac{25}{16}-1\right)}$$

or 400/3 yds.

75. THE TRIANGULAR FENCE

It will be interesting to compare your solution with that of Euclid, the young geometer, who recognized this instantly as the problem of passing a line through a point and cutting two intersecting lines so as to form a triangle of a given perimeter. He accordingly ruled off on a sheet of paper two lines CA and CB meeting at an angle of 80° and marked on each line to convenient scale (see Fig. 159) points D and E 100 ft. from the intersection. He drew a circle tangent to

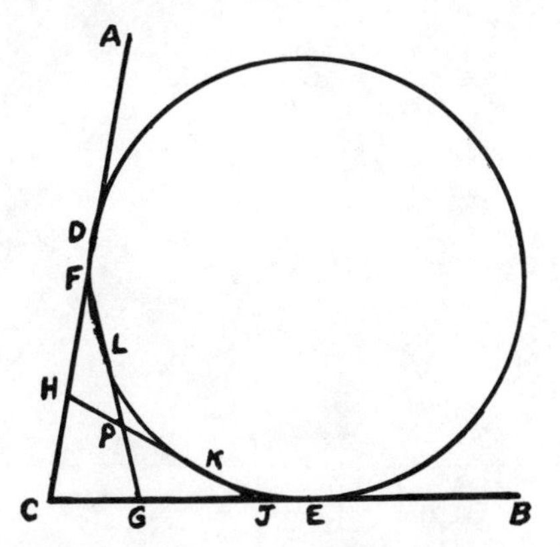

Fig. 159

the two lines at these points and then from the given point P he drew a tangent to the circle, there being two such tangents HJ and FG with points of tangency at K and L respectively. He scaled off the points where the tangents met the intersecting lines and the job was done.

For proof: HK equals HD and KJ equals JE, because tangents to a circle from a point are equal, from which HJ, when added to the two sides CH and CJ gives the required perimeter. Some of our readers, who forgot that Euclid was a geometer and ignored our hint about his doing the job "in two shakes", set up two simultaneous trigonometric equations and obtained the required answers numerically to several decimals.

SOLUTIONS TO QUICKIES

As the name implies, these problems are all capable of rapid solution if the right approach is used. In fact, most of them have been specially selected because they provide a test of skill in avoiding the roundabout road and finding the short cut to the answer. An approximate time for solution by the methods given here, is five minutes which you can compare with your own performance.

76. BALLS IN THE BOX

Of the great number of correct and varied solutions received to this Quickie when it appeared in the DIAL the most interesting was contributed by William S. Hornbaker, Oak Ridge National Laboratory, who realized that the dis-

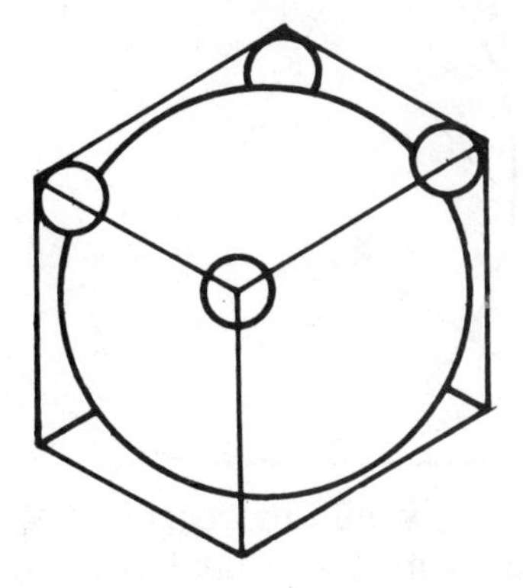

Fig. 160

tance from the center of the box (and of the large ball) to the corner of the box is equal to $R\sqrt{3}$. That distance is obviously equal to the square root of the sum of the squares of half the side of the cube and half a diagonal on one of the faces and for a similar reason the distance of the small ball center to the box corner is $r\sqrt{3}$, where r is the required radius. It is then evident from Fig. 160 that $R\sqrt{3} - R = r\sqrt{3} + r$ from which $r = (2 - \sqrt{3})R$ which when $R = 1$ as in this case gives the answer $r = .268''$. This analysis is of course, readily extended, wrote Mr. Hornbaker, to the placing of a succession of smaller balls in the pockets, the radius of the n th set of balls being $r_n = (2 - \sqrt{3})^n R$ which led Marvin Jacoby of Remington Rand, Philadelphia, who made a similar analysis, to the intriguing conclusion that

> Great balls in cubes have little balls
> In corners to delight them.
> And little balls have lesser balls,
> And so ad infinitum.

A few months after this Quickie appeared it happened that Dr. John W. Mauchly of Remington-Rand (now Sperry Rand), when told of the problem offered a cuter though no shorter solution by consideration of the infinite sequence of smaller and smaller balls that might be fitted into the pockets, which sequence prompted the quatrain above. The diam-

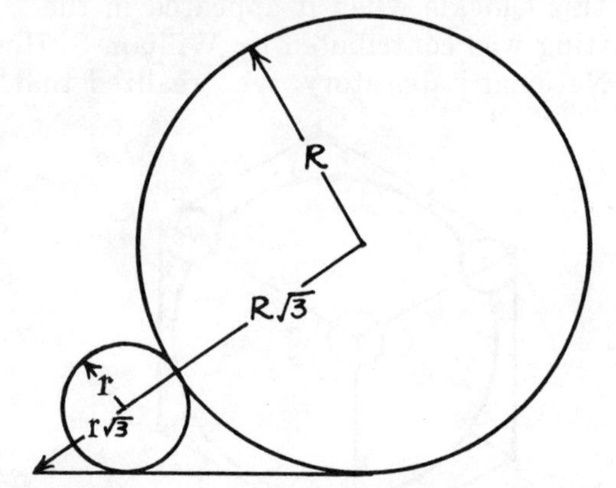

Fig. 161

eter of successive balls would by reason of symmetry have a constant fractional ratio c, so that if the largest ball had a diameter of unity; the infinite sum of the diameters of all balls would be $1/(1 - c)$, the diagonal $\sqrt{3}$ being equal to twice this amount less unity, which solves for c.

77. THE FOUR-CUSHION SHOT

Of the many possible modes of solution, the best and quickest is with mirrors, as in our problem No. 8, "From

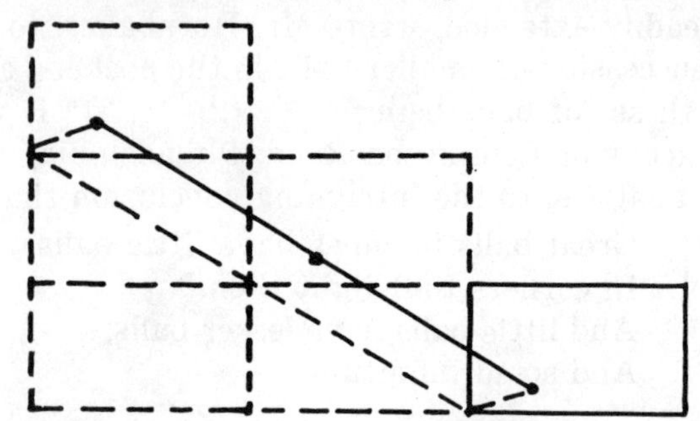

Fig. 162

Pole to Pole." Since the path of the billiard ball in mirrored tables (see Fig. 162) must be a straight line, it is obvious that for this line to terminate at the original point, it must be parallel to the diagonal and equal in length to two diagonals, since lines drawn from the ends of the path to the extremities of the diagonals would be equal and parallel and thus complete a parallelogram.

78. SUM OF THE CUBES

The short cut here is to divide both sides of the equation $a^3 + b^3 = c^4$ by c^3 which gives the relation $(a/c)^3 + (b/c)^3 = c$. If a/c and b/c are set equal to any two different figures other than 1, the numbers required are quickly found and are different integers. For example, if a/c and b/c are 2 and 3 respectively, c is $2^3 + 3^3 = 35$ and the required numbers are 70 and 105.

79. THE GRAZING COWS

This is an excellent example of how a great amount of time and effort can be saved and a seemingly involved problem turned into a simple mental exercise by avoiding the extraneous and making adroit use only of the facts required to reach the specific answer wanted. Although three different grazing situations are given, the first can be ignored since there are enough facts in the last two to furnish the answer in short order. This is because in each of these cases the original growth and new growth over the same 5-week period are consumed, which means that the total grass eaten is proportional to the acreage, so that the unknown number of cows eat 5/2 as much grass as the 8 cows, in 5/3 the time, requiring 3/2 as many cows or 12. The superfluous facts given establish the ratio between original growth per acre and rate of new growth etc., but this and other interesting relations are neither asked for nor necessary to the solution.

80. THE WOBBLY WHEEL

It is natural to assume that our mechanic was no mathematician, but did have the usual means for marking off angles on a circle, in this case on the circle of six-inch radius. So with the wheel shaft resting on the knife edges he would mark a point on the circle directly above the wheel center

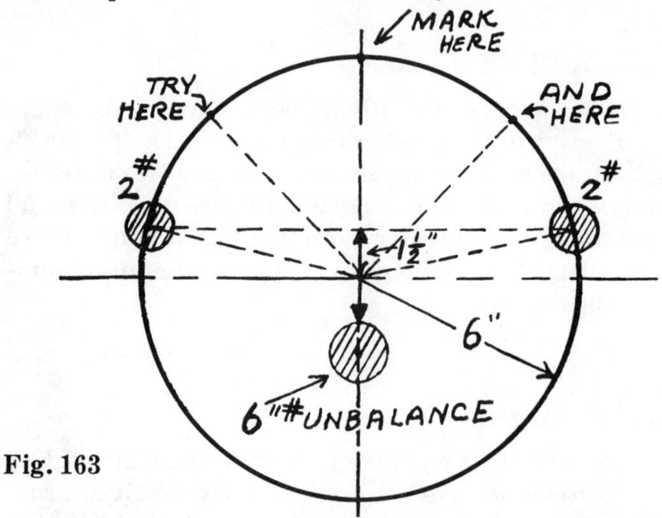

Fig. 163

(see Fig. 163). By trial and error he would then fasten each weight at various equal distances from that point, starting say at 45° each side of top and then moving the weights equally toward or away from each other, until the wheel stayed put in any position. If the weights were each two pounds and the unbalance 6 inch-pounds, the combined weights would have to correct that amount when the center of unbalance was horizontal; at which position each two pound weight must have an arm of 1-1/2″, the chord distance between them being $2\sqrt{36 - 9/4} = 11.5$ inches.

81. OLD MAN AND HIS SON

Clearly, each traveler when mounting the horse must have just walked the distance that his companion had previously chosen to ride. If they met at the 32-mile half-way point and the father had by then walked, in no matter what stages, a total distance W; the equation for the elapsed time would be $W/3 + (32 - W)/8 = W/8 + (32 - W)/4$ which gives $W = 12$ and the elapsed time 6.5 hours. Multiplying this by two, on the assumption that they will similarly meet at the destination, and adding the half-hour for lunch gives the arrival time of 7:30 P.M.

82. SHEET OF PAPER

To do the job by simple graphic means, the sheet of paper is drawn to convenient scale ABEH (Fig. 164). EB is extended to J so that BJ equals AB and a semi-circle drawn

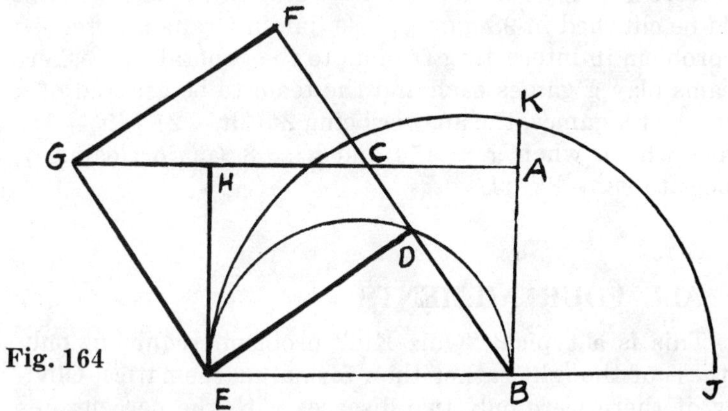

Fig. 164

EJ, BA then being extended to K. A semi-circle is drawn on EB and ED marked off on it equal to BK, BD joined and extended until it cuts AH at C. The three required pieces are triangles ABC and BED and quadrilateral CDEH. For proof draw EG parallel to BC and intersecting AH extended at G. Draw GF parallel to DE and intersecting BC extended at F, forming rectangle DEFG, which is readily shown to be a square comprising two triangles and a quadrilateral equal to the corresponding parts of the original rectangle.

83. EARLIEST TIE FOR THE PENNANT

A succinct answer to this problem in sportwriter's parlance was furnished by Roy Fink of Kent Plastics Corporation, Evansville, who wrote: "To clinch the championship at the earliest possible date, the winning team must win all its games and the other seven teams must all be tied for second place. This would require the also-rans to play .500 ball against each other while losing all their games to the champs. Therefore each of the also-rans would win 3/7 of its games while losing 4/7. The leading team has clinched a tie for the pennant when the number of its wins plus the number of its nearest rival's losses is equal to the number of games to be played (154), this being the method the sports writers use to calculate the so-called magic number. Therefore $x + 4/7x = 154$ and $x = 98$. A tie for the pennant could be clinched in 98 games." Benjamin Graham extended the problem in interesting fashion to the general case where h teams play g games each and the team to be assured of a tie has lost k games, the answer being $2/(3h - 2)\ [(h - 1)g + hk]$ which, when $g = 154$ and $h = 8$ (major leagues), reduces to $98 + 8k/11$.

84. ALL FOUR AILMENTS

This is a typical "Quiz Kid" problem, requiring only alertness of thought rather than formal mathematics. Obviously, if there were only two diseases with the percentages 70% and 75%, the minimum overlap, would be 45%, or 70% plus 75% minus 100. The minimum of 45% of the population having the first two diseases would similarly overlap the 80% with the third ailment by a minimum of 25%, and the minimum of 25% with the first three diseases would overlap the 85% with the fourth disease by at least 10%. The same handling would of course apply to any other combination of any number of ailments. The answer can be had instantly by subtracting from the total of all percentages given, a figure equal to 100 times one less than the number of ailments.

85. DELETED CHECKERBOARD

Many varieties of solution to this Quickie are possible by consideration of the axis of the domino or other means, but the pithiest answer received when this problem appeared in the DIAL was, in our opinion, the contribution by Walter

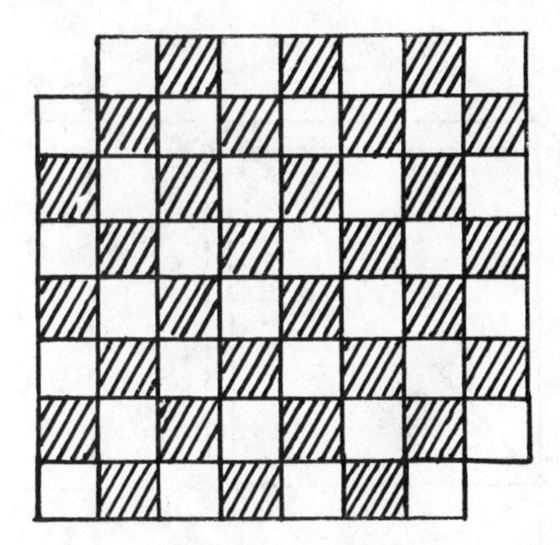

Fig. 165

P. Targoff, Eggertsville, N. Y.: "Inasmuch as the diagonally opposite corners of a checkerboard are of the same color, and as any one of the dominoes must cover one black and one white square; since adjacent squares are in all cases of different color, it is obviously impossible to cover 30 squares of one color plus 32 of the other color with 31 dominoes."

86. THE PLATE OF PIE

Wm. S. Hornbaker, Engineer, Oak Ridge National Laboratory, completed this solution in five minutes: "Consider a semi-circular segment of pie of unit radius whose area is $\pi/2$ or 1.5708. This segment is cut into a large, even number of very small sectors which are paired to approximate narrow rectangles of unit length as shown below. If the number of segments is increased, the arc decreases, and ultimately the paired segments approach a true narrow rectangle of unit length. Arranging the rectangle as shown in the sketch, we have a cross whose center is a square of 1 square unit, and the four rectangles are 1 unit long by 1/4 of (1.5708 − 1) or 0.1427 units wide. From Fig. 166 it is

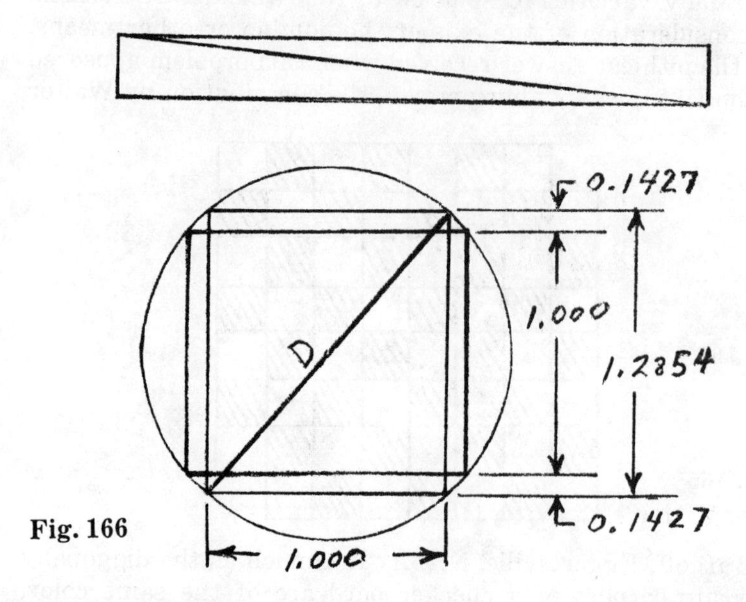

Fig. 166

easily seen that the diameter of the required circle equals $\sqrt{1 + (1.2854)^2}$ or 1.628, which means that if the half-pie can be cut into thin enough slivers, it can be disposed on a plate with radius 81.4% of that of the original dish, said smaller plate having an unused area of only 24.5%." Another DIAL reader who used this same method, Harry S. Reizenstein, Ordnance Design Engineer, Westinghouse, remarked that this Quickie made him "think of the optimum core shape for transformers with cylindrical coils, which is usually cruciform."

87. ROD IN THE BEAKER

A short and straightforward answer to this Quickie is: Two cubic centimeters of water must be added to the second beaker; one to balance the one c.c. of water added to the first beaker, and one to balance the buoyant force exerted on the iron bar due to its displacing one c.c. of water when the water level has been raised one centimeter.

Fig. 167

88. MARKET FOR HOGS

Vic Yingling of Harris-Seybold Company, Cleveland, set the problem down neatly, as below:

$$(\text{Man})^2 - (\text{Wife})^2 = 63$$
$$32^2 - 31^2 = 63$$
$$12^2 - 9^2 = 63$$
$$8^2 - 1^2 = 63$$

John 32	James 12
−23	−11
Sue 9	Mary 1

So it's: John-Ann
James-Sue
Henry-Mary

There are only three possible solutions of the equation $X^2 - Y^2 = 63$ as given above, because $(X + Y)$ must be a factor of 63, and so must the smaller number $(X - Y)$, from which $X - Y$ must be either 7, 3, or 1, the corresponding values of $X + Y$ being 9, 21, and 63, which of course determine X and Y.

89. PASSING STREETCARS

The conventional solution to this problem by the method of relative speeds is had by stating that the relative velocity between man and car when going respectively in the same and opposite directions is proportional to the number of cars encountered, which establish the equation $(x + 3)/(x - 3) = 60/40$ from which $x = 15$ mph. However, a less technical and perhaps more lucid explanation that short-circuits the algebra was furnished by Loyd H. Jones, G. E. Company, Syracuse. Mr. Jones pictured two cars at the start of the walk, the 40th car behind the man and the 60th car ahead of him. These must obviously have each traveled half the distance between them when they met at the man, namely, a 50-car space, so that the distance walked in the same period was a 10-car space or one-fifth as much, which means that the car speed was 15 mph.

90. COMMON BIRTHDAYS

Since there are $(365)^{30}$ possible choices for 30 birthdays, of which 364 x 363 x 362 - - - x 336 are without duplications, the quotient of the two gives the odds that none of the 30 boys would have identical birthdays, which figures out to be .294. If Euclid made a 10¢ to 50¢ bet in each of the 24 classrooms he would (with average luck) lose in seven cases and win in seventeen cases. His losses would be 70¢ and his gains $8.50, leaving a neat profit of $7.80. Incidentally, Mr. Phelps, who authored the problem, reminds us that two of our Presidents; Polk and Harding, had the same birthday, November 2nd, and two others; John Adams and Jefferson, the same date of death, July 4. This provides interesting confirmation of the fact that in any group of that size there's much more than an even chance that at least two will have significant happenings on identical days.

91. RESISTANCE 'CROSS THE CUBE

This, like the "Walk Around" problem, No. 66, is readily solved if one takes the easy road of symmetry. Of the many correct solutions that follow that path, perhaps the most interesting is one supplied by Edward Partridge of Allen B. Dumont Lab., Passaic, N. J., who provided the helpful dia-

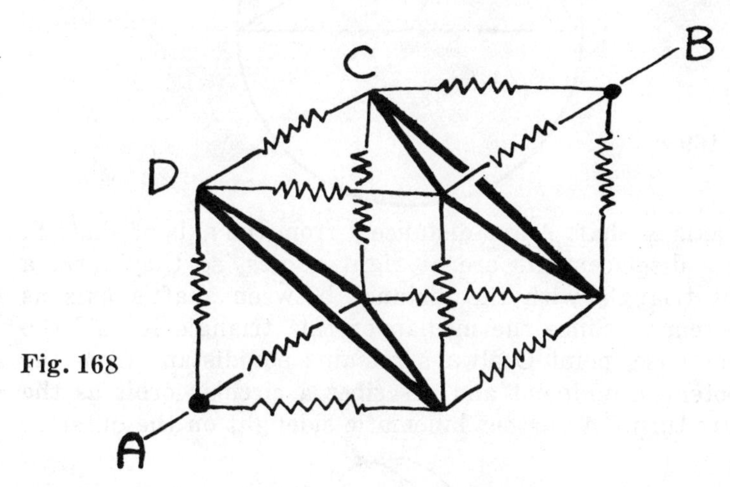

Fig. 168

gram (Fig. 168) and wrote: "Short the three corners nearest to corner A together, and the three corners nearest to corner B together. This can be done without disturbing the current flow since all three of each set of shorted points are equipotential (or symmetrical) points, and the shorts will carry no current. Then there are three equal resistors in parallel between points A and D, six between D and C, and three between C and B; the required resistance being $1/3 + 1/6 + 1/3$ or $5/6$ ohm."

92. MEDIANS AND OLDHAMS

Wrote Mr. R. Harrington, B. F. Goodrich Research Center, Brecksville, Ohio: "Draw a circle having the hypotenuse of the triangle as a diameter: then the apex must fall on the circle. The median of the problem is a radius and is thus half the hypotenuse (see Fig. 169). In an Oldham coupling the center C of the middle element is displaced distance x from

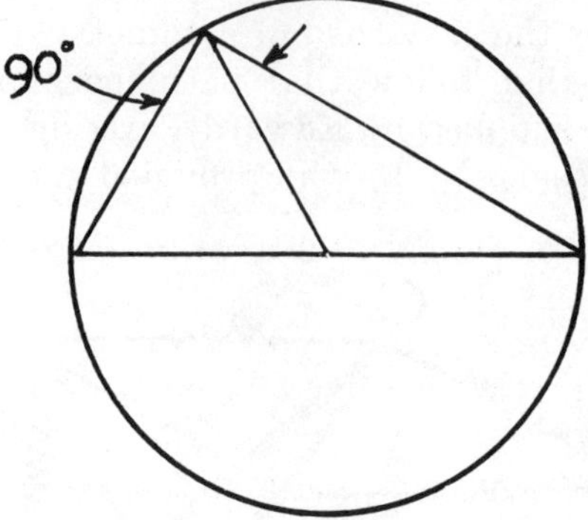

Fig. 169

the axis of shaft A and distance y from the axis of shaft B. These displacements are at right angles, so they form a right triangle with the distance between shaft's axis as hypotenuse. Since the median of this triangle is half the hypotenuse, point C always remains equidistant from the hypotenuse midpoint and describes a circular orbit as the shafts turn." A further kinematic sidelight on the question

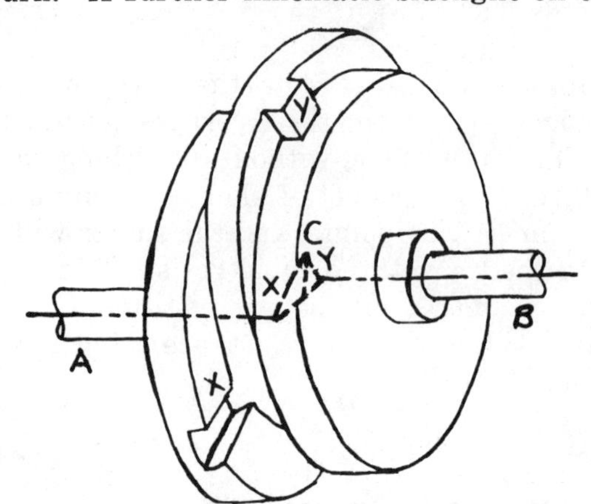

Fig. 170

is that the Oldham coupling (Fig. 170) is an inversion of the familiar isosceles linkage wherein a crank, connecting rod and slider mechanism with crank and connecting rod equal in length (and so forming an isosceles triangle) can be replaced by a single rod with slider at each end, moving in mutually perpendicular guides because of the relationship stated in the Quickie.

93. THE BACKWARD BICYCLE

A particularly enlightening solution to this problem read as follows: In order for an object to move when a force is applied, it is imperative that the point at which the force is applied is free to move in the direction of the force. It is obvious that if a force is applied to a point on a spoke in the rear wheel, the wheel will roll in the direction of the force. If the point is between the axle and the ground, the point moves forward with respect to the axle but backward with respect to the ground and in relation to the force moves in its direction. Due to the small velocity ratio (ratio of RPM of pedal sprocket to RPM of wheel sprocket), inherent in bicycles in order to attain speed, the pedal will move slowly forward with respect to the frame but its motion is backward with respect to the ground and the applied force, hence the bicycle rolls backward. If the force had been applied to the upper pedal, the brakes would set and the answer is (c). When this problem appeared in the DIAL two Californians were inspired to break into doggerel, Robert Chamberlain of Los Altos:

Solving this problem took little time,
But composing this poem took much more of mine,
However, the answer is not very hard
If you do as I did, just go out in the yard,
You borrow a bike and hold it with love,
Then set up the pedals and give them a shove,
When this is all done, I'm sure you will find
The gearing is such that the vector's behind.
Part of the force attempts to go fore,
But all of the force accomplished the chore.

and Lowell Sisson of Whittier:

The prime principle of a bicycle is based
On the human desire to make haste.

Hence the tire passes over the earth more fleet
Than the pedals on which rest the feet.
So if speed by the ratio is gained
Then the force must be equally drained,
And the rearward force on the step
Will be met by one which is less
From the wheel which is in draft,
Thus the vehicle will then move aft.

And an interesting comment was made by Ralph Bowersox, of Los Angeles, who reminded us that, "The problem is that of a rolling wheel geared to the ground, in which it is an elementary but to some people surprising fact that no point in or on the wheel ever moves in a direction contrary to the center of the wheel." (As illustrated in the space-time curves of a cycloid or prolate trochoid.)

94. EVEN NUMBER OF DOORS

R. V. Gillespie, Duff Norton Company, Pittsburgh, reasoned in this fashion: "Each external door of the house has one side facing out, so we must show that there are an even number of door sides on the outside. If the house has N doors altogether, it has 2N door sides. The problem says essentially that each room of the house has an even number of door sides facing into it. Thus there are an even number, say 2K, of door sides on the inside of the house. The number of door sides on the outside is then $2N - 2K =$ an even number."

95. THE AMBULATORY COMMUTER

It is obvious that if the Grahams got home 15 minutes earlier than usual, Mrs. Graham must have saved two 7-1/2-minute driving stretches to and from the station of 3-3/4 miles each at her rate of 30 mph. Since she was due here at 5 P.M., she must have met her spouse at 4:52.5, after he had walked the 3-3/4 miles in 52-1/2 minutes, or at a rate of

4-2/7 mph. This makes the thing pretty much a mental exercise since the fractions cancel out requiring only a few minutes to reach the answer, but many much more lengthy and circuitous modes of solution are possible. Some of these superfluously involve the distance from home to station, which obviously must be less than 30 miles (and from a practical angle is doubtless less than 5 miles) but does not affect the answer.

96. G.C.D. AND L.C.M.

Let any two numbers be N and M. By definition of G.C.D. designated here as G, N = GA and M = GB, where A and B are integers having no common factors. The L.C.M. of M and N is GAB and their sum is G(A + B). Since A and B have no common factors, (A + B) likewise has no common factors with A or B, or AB. Therefore G is also the G.C.D. of the L.C.M. and the sum of N and M. This can be illustrated by two numbers taken at random, 42 and 144, which are written 6 x 7 and 6 x 24, whereby G is 6, A is 7, and B is 24. The L.C.M. is GAB, or 6 x 7 x 24 = 1048. The sum of 42 and 144, or 186, is equal to 6(7 + 24) and since 7 and 24 have no common factor, neither can have a common factor with their sum.

97. THE TRACKWALKERS' QUANDARY

A conventional type of solution using familiar methods of relative velocity reads: Since it takes the train of any length "L" 10 seconds to pass A, and 9 seconds to pass B, the relative velocity between train and A is L/10, between train and B is L/9, and between A and B is L/90; the latter figure being one-tenth the relative velocity between train and B. Since it took the train 1210 seconds to reach B, it will take A ten times as long, or 12,100 seconds, of which 1219 seconds had elapsed when the train passed B, leaving 3 hours, 1 minute, and 21 seconds. This is a fairly simple and rapid approach and avoids the quite correct but unnecessarily roundabout method which introduces "let x be this and y that," and brings in da/dt or other differentials.

An alternative and perhaps still more interesting solution considers an observer looking out of a supposedly stationary train at the two trackwalkers successively whizzing by. "It appears to such an observer," wrote Mr. Alexander Yorgiadis, "that the second man moves faster than the first, since the second man takes nine seconds to cover a distance that the first man covers in ten seconds, and thus in a nine second period, the second man gains one second over the first. It is given that the second man goes past the rear end of the train twenty minutes and nine seconds after the first man, and for them to meet it would take nine times this interval, or three hours, one minute, and twenty-one seconds."

98. CONSECUTIVE INTEGERS

When this problem appeared in the DIAL many readers "instinctively" hit on the first integral sequence 1, 2, 3; without aid of algebra; the fractions $1/2 + 1/3 + 2/3 + 2/1 + 3/2 + 3/1$ adding up to 8. However, it required only a few moments more to show adroitly that this was actually the only sequence of positive numbers that could fill the bill. Setting the thing up algebraically, $x - 1, x, x + 1$ to give the sum of $6x^2/(x^2 - 1)$, it is readily seen that since x^2 and $(x^2 - 1)$ are relatively prime, $(x^2 - 1)$ must be a divisor of 6, which is possible only if $x = 2$.

99. COWS, HORSES, AND CHICKENS

The equation $C(C + H) = 120 + F$ where F means fowl, reveals that F cannot equal 2, the only even prime number, since C would then also have to be 2, which is not a different number, and $C + H = 61$, so F must be odd, from which either C or H must equal 2, since the sum of the two would otherwise be even. Clearly, H must then be 2, since C equals 2 would make the product even. We then rewrite the equation $C^2 + 2C - 120 = F$ or $(C + 12)(C - 10) = F$. $C - 10$ must equal unity in order for F to be prime, from which $C = 11$ and $F = 23$, the only possible solution.

100. BINGO CARDS

It is readily possible, by failing to analyze the situation clearly, to make this Quickie more complicated than it actually is. Clearly, the number of cards satisfying the requirement that no two may be in an identical row, column or diagonal cannot exceed the number of cards with different middle columns. This is, in fact, the total number of possible different cards since the number of cards with different rows and diagonals is greater than this number $C(15, 4) = 1365$.

A CATALOGUE OF SELECTED DOVER BOOKS IN ALL FIELDS OF INTEREST

A CATALOGUE OF SELECTED DOVER BOOKS IN ALL FIELDS OF INTEREST

THE SENSE OF BEAUTY, George Santayana. Masterfully written discussion of nature of beauty, materials of beauty, form, expression; art, literature, social sciences all involved. 168pp. 5⅜ x 8½. 20238-0 Pa. $3.50

ON THE IMPROVEMENT OF THE UNDERSTANDING, Benedict Spinoza. Also contains *Ethics, Correspondence,* all in excellent R. Elwes translation. Basic works on entry to philosophy, pantheism, exchange of ideas with great contemporaries. 402pp. 5⅜ x 8½. 20250-X Pa. $5.95

THE TRAGIC SENSE OF LIFE, Miguel de Unamuno. Acknowledged masterpiece of existential literature, one of most important books of 20th century. Introduction by Madariaga. 367pp. 5⅜ x 8½.
20257-7 Pa. $6.00

THE GUIDE FOR THE PERPLEXED, Moses Maimonides. Great classic of medieval Judaism attempts to reconcile revealed religion (Pentateuch, commentaries) with Aristotelian philosophy. Important historically, still relevant in problems. Unabridged Friedlander translation. Total of 473pp. 5⅜ x 8½. 20351-4 Pa. $6.95

THE I CHING (THE BOOK OF CHANGES), translated by James Legge. Complete translation of basic text plus appendices by Confucius, and Chinese commentary of most penetrating divination manual ever prepared. Indispensable to study of early Oriental civilizations, to modern inquiring reader. 448pp. 5⅜ x 8½. 21062-6 Pa. $6.00

THE EGYPTIAN BOOK OF THE DEAD, E. A. Wallis Budge. Complete reproduction of Ani's papyrus, finest ever found. Full hieroglyphic text, interlinear transliteration, word for word translation, smooth translation. Basic work, for Egyptology, for modern study of psychic matters. Total of 533pp. 6½ x 9¼. (USCO) 21866-X Pa. $8.50

THE GODS OF THE EGYPTIANS, E. A. Wallis Budge. Never excelled for richness, fullness: all gods, goddesses, demons, mythical figures of Ancient Egypt; their legends, rites, incarnations, variations, powers, etc. Many hieroglyphic texts cited. Over 225 illustrations, plus 6 color plates. Total of 988pp. 6⅛ x 9¼. (EBE)
22055-9, 22056-7 Pa., Two-vol. set $20.00

THE STANDARD BOOK OF QUILT MAKING AND COLLECTING, Marguerite Ickis. Full information, full-sized patterns for making 46 traditional quilts, also 150 other patterns. Quilted cloths, lame, satin quilts, etc. 483 illustrations. 273pp. 6⅞ x 9⅝. 20582-7 Pa. $5.95

CORAL GARDENS AND THEIR MAGIC, Bronsilaw Malinowski. Classic study of the methods of tilling the soil and of agricultural rites in the Trobriand Islands of Melanesia. Author is one of the most important figures in the field of modern social anthropology. 143 illustrations. Indexes. Total of 911pp. of text. 5⅝ x 8¼. (Available in U.S. only)
23597-1 Pa. $12.95

THE PHILOSOPHY OF HISTORY, Georg W. Hegel. Great classic of Western thought develops concept that history is not chance but a rational process, the evolution of freedom. 457pp. 5⅜ x 8½. **20112-0 Pa. $6.00**

LANGUAGE, TRUTH AND LOGIC, Alfred J. Ayer. Famous, clear introduction to Vienna, Cambridge schools of Logical Positivism. Role of philosophy, elimination of metaphysics, nature of analysis, etc. 160pp. 5⅜ x 8½. (USCO) 20010-8 Pa. $2.50

A PREFACE TO LOGIC, Morris R. Cohen. Great City College teacher in renowned, easily followed exposition of formal logic, probability, values, logic and world order and similar topics; no previous background needed. 209pp. 5⅜ x 8½. 23517-3 Pa. $4.95

REASON AND NATURE, Morris R. Cohen. Brilliant analysis of reason and its multitudinous ramifications by charismatic teacher. Interdisciplinary, synthesizing work widely praised when it first appeared in 1931. Second (1953) edition. Indexes. 496pp. 5⅜ x 8½. 23633-1 Pa. $7.50

AN ESSAY CONCERNING HUMAN UNDERSTANDING, John Locke. The only complete edition of enormously important classic, with authoritative editorial material by A. C. Fraser. Total of 1176pp. 5⅜ x 8½. 20530-4, 20531-2 Pa., Two-vol. set $16.00

HANDBOOK OF MATHEMATICAL FUNCTIONS WITH FORMULAS, GRAPHS, AND MATHEMATICAL TABLES, edited by Milton Abramowitz and Irene A. Stegun. Vast compendium: 29 sets of tables, some to as high as 20 places. 1,046pp. 8 x 10½. 61272-4 Pa. $17.95

MATHEMATICS FOR THE PHYSICAL SCIENCES, Herbert S. Wilf. Highly acclaimed work offers clear presentations of vector spaces and matrices, orthogonal functions, roots of polynomial equations, conformal mapping, calculus of variations, etc. Knowledge of theory of functions of real and complex variables is assumed. Exercises and solutions. Index. 284pp. 5⅝ x 8¼. 63635-6 Pa. $5.00

THE PRINCIPLE OF RELATIVITY, Albert Einstein et al. Eleven most important original papers on special and general theories. Seven by Einstein, two by Lorentz, one each by Minkowski and Weyl. All translated, unabridged. 216pp. 5⅜ x 8½. 60081-5 Pa. $3.50

THERMODYNAMICS, Enrico Fermi. A classic of modern science. Clear, organized treatment of systems, first and second laws, entropy, thermodynamic potentials, gaseous reactions, dilute solutions, entropy constant. No math beyond calculus required. Problems. 160pp. 5⅜ x 8½. 60361-X Pa. $4.00

ELEMENTARY MECHANICS OF FLUIDS, Hunter Rouse. Classic undergraduate text widely considered to be far better than many later books. Ranges from fluid velocity and acceleration to role of compressibility in fluid motion. Numerous examples, questions, problems. 224 illustrations. 376pp. 5⅝ x 8¼. 63699-2 Pa. $7.00

THE AMERICAN SENATOR, Anthony Trollope. Little known, long unavailable Trollope novel on a grand scale. Here are humorous comment on American vs. English culture, and stunning portrayal of a heroine/villainess. Superb evocation of Victorian village life. 561pp. 5⅜ x 8½.
23801-6 Pa. $7.95

WAS IT MURDER? James Hilton. The author of *Lost Horizon* and *Goodbye, Mr. Chips* wrote one detective novel (under a pen-name) which was quickly forgotten and virtually lost, even at the height of Hilton's fame. This edition brings it back—a finely crafted public school puzzle resplendent with Hilton's stylish atmosphere. A thoroughly English thriller by the creator of Shangri-la. 252pp. 5⅜ x 8. (Available in U.S. only)
23774-5 Pa. $3.00

CENTRAL PARK: A PHOTOGRAPHIC GUIDE, Victor Laredo and Henry Hope Reed. 121 superb photographs show dramatic views of Central Park: Bethesda Fountain, Cleopatra's Needle, Sheep Meadow, the Blockhouse, plus people engaged in many park activities: ice skating, bike riding, etc. Captions by former Curator of Central Park, Henry Hope Reed, provide historical view, changes, etc. Also photos of N.Y. landmarks on park's periphery. 96pp. 8½ x 11. 23750-8 Pa. $4.50

NANTUCKET IN THE NINETEENTH CENTURY, Clay Lancaster. 180 rare photographs, stereographs, maps, drawings and floor plans recreate unique American island society. Authentic scenes of shipwreck, lighthouses, streets, homes are arranged in geographic sequence to provide walking-tour guide to old Nantucket existing today. Introduction, captions. 160pp. 8⅞ x 11¾. 23747-8 Pa. $7.95

STONE AND MAN: A PHOTOGRAPHIC EXPLORATION, Andreas Feininger. 106 photographs by *Life* photographer Feininger portray man's deep passion for stone through the ages. Stonehenge-like megaliths, fortified towns, sculpted marble and crumbling tenements show textures, beauties, fascination. 128pp. 9¼ x 10¾. 23756-7 Pa. $5.95

CIRCLES, A MATHEMATICAL VIEW, D. Pedoe. Fundamental aspects of college geometry, non-Euclidean geometry, and other branches of mathematics: representing circle by point. Poincare model, isoperimetric property, etc. Stimulating recreational reading. 66 figures. 96pp. 5⅝ x 8¼.
63698-4 Pa. $3.50

THE DISCOVERY OF NEPTUNE, Morton Grosser. Dramatic scientific history of the investigations leading up to the actual discovery of the eighth planet of our solar system. Lucid, well-researched book by well-known historian of science. 172pp. 5⅜ x 8½. 23726-5 Pa. $3.50

THE DEVIL'S DICTIONARY. Ambrose Bierce. Barbed, bitter, brilliant witticisms in the form of a dictionary. Best, most ferocious satire America has produced. 145pp. 5⅜ x 8½. 20487-1 Pa. $2.50

CATALOGUE OF DOVER BOOKS

HISTORY OF BACTERIOLOGY, William Bulloch. The only comprehensive history of bacteriology from the beginnings through the 19th century. Special emphasis is given to biography-Leeuwenhoek, etc. Brief accounts of 350 bacteriologists form a separate section. No clearer, fuller study, suitable to scientists and general readers, has yet been written. 52 illustrations. 448pp. 5⅝ x 8¼. 23761-3 Pa. $6.50

THE COMPLETE NONSENSE OF EDWARD LEAR, Edward Lear. All nonsense limericks, zany alphabets, Owl and Pussycat, songs, nonsense botany, etc., illustrated by Lear. Total of 321pp. 5⅜ x 8½. (Available in U.S. only) 20167-8 Pa. $4.50

INGENIOUS MATHEMATICAL PROBLEMS AND METHODS, Louis A. Graham. Sophisticated material from Graham *Dial*, applied and pure; stresses solution methods. Logic, number theory, networks, inversions, etc. 237pp. 5⅜ x 8½. 20545-2 Pa. $4.50

BEST MATHEMATICAL PUZZLES OF SAM LOYD, edited by Martin Gardner. Bizarre, original, whimsical puzzles by America's greatest puzzler. From fabulously rare *Cyclopedia*, including famous 14-15 puzzles, the Horse of a Different Color, 115 more. Elementary math. 150 illustrations. 167pp. 5⅜ x 8½. 20498-7 Pa. $3.50

THE BASIS OF COMBINATION IN CHESS, J. du Mont. Easy-to-follow, instructive book on elements of combination play, with chapters on each piece and every powerful combination team—two knights, bishop and knight, rook and bishop, etc. 250 diagrams. 218pp. 5⅜ x 8½. (Available in U.S. only) 23644-7 Pa. $4.50

MODERN CHESS STRATEGY, Ludek Pachman. The use of the queen, the active king, exchanges, pawn play, the center, weak squares, etc. Section on rook alone worth price of the book. Stress on the moderns. Often considered the most important book on strategy. 314pp. 5⅜ x 8½. 20290-9 Pa. $5.00

LASKER'S MANUAL OF CHESS, Dr. Emanuel Lasker. Great world champion offers very thorough coverage of all aspects of chess. Combinations, position play, openings, end game, aesthetics of chess, philosophy of struggle, much more. Filled with analyzed games. 390pp. 5⅜ x 8½. 20640-8 Pa. $5.95

500 MASTER GAMES OF CHESS, S. Tartakower, J. du Mont. Vast collection of great chess games from 1798-1938, with much material nowhere else readily available. Fully annoted, arranged by opening for easier study. 664pp. 5⅜ x 8½. 23208-5 Pa. $8.50

A GUIDE TO CHESS ENDINGS, Dr. Max Euwe, David Hooper. One of the finest modern works on chess endings. Thorough analysis of the most frequently encountered endings by former world champion. 331 examples, each with diagram. 248pp. 5⅜ x 8½. 23332-4 Pa. $3.95

THE COMPLETE BOOK OF DOLL MAKING AND COLLECTING, Catherine Christopher. Instructions, patterns for dozens of dolls, from rag doll on up to elaborate, historically accurate figures. Mould faces, sew clothing, make doll houses, etc. Also collecting information. Many illustrations. 288pp. 6 x 9. 22066-4 Pa. $4.95

THE DAGUERREOTYPE IN AMERICA, Beaumont Newhall. Wonderful portraits, 1850's townscapes, landscapes; full text plus 104 photographs. The basic book. Enlarged 1976 edition. 272pp. 8¼ x 11¼. 23322-7 Pa. $7.95

CRAFTSMAN HOMES, Gustav Stickley. 296 architectural drawings, floor plans, and photographs illustrate 40 different kinds of "Mission-style" homes from *The Craftsman* (1901-16), voice of American style of simplicity and organic harmony. Thorough coverage of Craftsman idea in text and picture, now collector's item. 224pp. 8⅛ x 11. 23791-5 Pa. $6.50

PEWTER-WORKING: INSTRUCTIONS AND PROJECTS, Burl N. Osborn. & Gordon O. Wilber. Introduction to pewter-working for amateur craftsman. History and characteristics of pewter; tools, materials, step-by-step instructions. Photos, line drawings, diagrams. Total of 160pp. 7⅞ x 10¾. 23786-9 Pa. $3.50

THE GREAT CHICAGO FIRE, edited by David Lowe. 10 dramatic, eyewitness accounts of the 1871 disaster, including one of the aftermath and rebuilding, plus 70 contemporary photographs and illustrations of the ruins—courthouse, Palmer House, Great Central Depot, etc. Introduction by David Lowe. 87pp. 8¼ x 11. 23771-0 Pa. $4.00

SILHOUETTES: A PICTORIAL ARCHIVE OF VARIED ILLUSTRATIONS, edited by Carol Belanger Grafton. Over 600 silhouettes from the 18th to 20th centuries include profiles and full figures of men and women, children, birds and animals, groups and scenes, nature, ships, an alphabet. Dozens of uses for commercial artists and craftspeople. 144pp. 8⅜ x 11¼. 23781-8 Pa. $4.50

ANIMALS: 1,419 COPYRIGHT-FREE ILLUSTRATIONS OF MAMMALS, BIRDS, FISH, INSECTS, ETC., edited by Jim Harter. Clear wood engravings present, in extremely lifelike poses, over 1,000 species of animals. One of the most extensive copyright-free pictorial sourcebooks of its kind. Captions. Index. 284pp. 9 x 12. 23766-4 Pa. $8.95

INDIAN DESIGNS FROM ANCIENT ECUADOR, Frederick W. Shaffer. 282 original designs by pre-Columbian Indians of Ecuador (500-1500 A.D.). Designs include people, mammals, birds, reptiles, fish, plants, heads, geometric designs. Use as is or alter for advertising, textiles, leathercraft, etc. Introduction. 95pp. 8¾ x 11¼. 23764-8 Pa. $4.50

SZIGETI ON THE VIOLIN, Joseph Szigeti. Genial, loosely structured tour by premier violinist, featuring a pleasant mixture of reminiscenes, insights into great music and musicians, innumerable tips for practicing violinists. 385 musical passages. 256pp. 5⅝ x 8¼. 23763-X Pa. $4.00

TONE POEMS, SERIES II: TILL EULENSPIEGELS LUSTIGE STREICHE, ALSO SPRACH ZARATHUSTRA, AND EIN HELDENLEBEN, Richard Strauss. Three important orchestral works, including very popular *Till Eulenspiegel's Marry Pranks,* reproduced in full score from original editions. Study score. 315pp. 9⅜ x 12¼. (Available in U.S. only) 23755-9 Pa. $8.95

TONE POEMS, SERIES I: DON JUAN, TOD UND VERKLARUNG AND DON QUIXOTE, Richard Strauss. Three of the most often performed and recorded works in entire orchestral repertoire, reproduced in full score from original editions. Study score. 286pp. 9⅜ x 12¼. (Available in U.S. only) 23754-0 Pa. $8.95

11 LATE STRING QUARTETS, Franz Joseph Haydn. The form which Haydn defined and "brought to perfection." *(Grove's).* 11 string quartets in complete score, his last and his best. The first in a projected series of the complete Haydn string quartets. Reliable modern Eulenberg edition, otherwise difficult to obtain. 320pp. 8⅜ x 11¼. (Available in U.S. only) 23753-2 Pa. $8.95

FOURTH, FIFTH AND SIXTH SYMPHONIES IN FULL SCORE, Peter Ilyitch Tchaikovsky. Complete orchestral scores of Symphony No. 4 in F Minor, Op. 36; Symphony No. 5 in E Minor, Op. 64; Symphony No. 6 in B Minor, "Pathetique," Op. 74. Bretikopf & Hartel eds. Study score. 480pp. 9⅜ x 12¼. 23861-X Pa. $10.95

THE MARRIAGE OF FIGARO: COMPLETE SCORE, Wolfgang A. Mozart. Finest comic opera ever written. Full score, not to be confused with piano renderings. Peters edition. Study score. 448pp. 9⅜ x 12¼. (Available in U.S. only) 23751-6 Pa. $12.95

"IMAGE" ON THE ART AND EVOLUTION OF THE FILM, edited by Marshall Deutelbaum. Pioneering book brings together for first time 38 groundbreaking articles on early silent films from *Image* and 263 illustrations newly shot from rare prints in the collection of the International Museum of Photography. A landmark work. Index. 256pp. 8¼ x 11. 23777-X Pa. $8.95

AROUND-THE-WORLD COOKY BOOK, Lois Lintner Sumption and Marguerite Lintner Ashbrook. 373 cooky and frosting recipes from 28 countries (America, Austria, China, Russia, Italy, etc.) include Viennese kisses, rice wafers, London strips, lady fingers, hony, sugar spice, maple cookies, etc. Clear instructions. All tested. 38 drawings. 182pp. 5⅜ x 8. 23802-4 Pa. $2.75

THE ART NOUVEAU STYLE, edited by Roberta Waddell. 579 rare photographs, not available elsewhere, of works in jewelry, metalwork, glass, ceramics, textiles, architecture and furniture by 175 artists—Mucha, Seguy, Lalique, Tiffany, Gaudin, Hohlwein, Saarinen, and many others. 288pp. 8⅜ x 11¼. 23515-7 Pa. $8.95

GREAT NEWS PHOTOS AND THE STORIES BEHIND THEM, John Faber. Dramatic volume of 140 great news photos, 1855 through 1976, and revealing stories behind them, with both historical and technical information. Hindenburg disaster, shooting of Oswald, nomination of Jimmy Carter, etc. 160pp. 8¼ x 11. 23667-6 Pa. **$6.00**

CRUICKSHANK'S PHOTOGRAPHS OF BIRDS OF AMERICA, Allan D. Cruickshank. Great ornithologist, photographer presents 177 closeups, groupings, panoramas, flightings, etc., of about 150 different birds. Expanded *Wings in the Wilderness*. Introduction by Helen G. Cruickshank. 191pp. 8¼ x 11. 23497-5 Pa. $7.95

AMERICAN WILDLIFE AND PLANTS, A. C. Martin, et al. Describes food habits of more than 1000 species of mammals, birds, fish. Special treatment of important food plants. Over 300 illustrations. 500pp. 5⅜ x 8½. 20793-5 Pa. **$6.50**

THE PEOPLE CALLED SHAKERS, Edward D. Andrews. Lifetime of research, definitive study of Shakers: origins, beliefs, practices, dances, social organization, furniture and crafts, impact on 19th-century USA, present heritage. Indispensable to student of American history, collector. 33 illustrations. 351pp. 5⅜ x 8½. 21081-2 Pa. $4.50

OLD NEW YORK IN EARLY PHOTOGRAPHS, Mary Black. New York City as it was in 1853-1901, through 196 wonderful photographs from N.-Y. Historical Society. Great Blizzard, Lincoln's funeral procession, great buildings. 228pp. 9 x 12. 22907-6 Pa. $8.95

MR. LINCOLN'S CAMERA MAN: MATHEW BRADY, Roy Meredith. Over 300 Brady photos reproduced directly from original negatives, photos. Jackson, Webster, Grant, Lee, Carnegie, Barnum; Lincoln; Battle Smoke, Death of Rebel Sniper, Atlanta Just After Capture. Lively commentary. 368pp. 8⅜ x 11¼. 23021-X Pa. $11.95

TRAVELS OF WILLIAM BARTRAM, William Bartram. From 1773-8, Bartram explored Northern Florida, Georgia, Carolinas, and reported on wild life, plants, Indians, early settlers. Basic account for period, entertaining reading. Edited by Mark Van Doren. 13 illustrations. 141pp. 5⅜ x 8½. 20013-2 Pa. $6.00

THE GENTLEMAN AND CABINET MAKER'S DIRECTOR, Thomas Chippendale. Full reprint, 1762 style book, most influential of all time; chairs, tables, sofas, mirrors, cabinets, etc. 200 plates, plus 24 photographs of surviving pieces. 249pp. 9⅞ x 12¾. 21601-2 Pa. $8.95

AMERICAN CARRIAGES, SLEIGHS, SULKIES AND CARTS, edited by Don H. Berkebile. 168 Victorian illustrations from catalogues, trade journals, fully captioned. Useful for artists. Author is Assoc. Curator, Div. of Transportation of Smithsonian Institution. 168pp. 8½ x 9½. 23328-6 Pa. $5.00

SECOND PIATIGORSKY CUP, edited by Isaac Kashdan. One of the greatest tournament books ever produced in the English language. All 90 games of the 1966 tournament, annotated by players, most annotated by both players. Features Petrosian, Spassky, Fischer, Larsen, six others. 228pp. 5⅜ x 8½. 23572-6 Pa. $3.50

ENCYCLOPEDIA OF CARD TRICKS, revised and edited by Jean Hugard. How to perform over 600 card tricks, devised by the world's greatest magicians: impromptus, spelling tricks, key cards, using special packs, much, much more. Additional chapter on card technique. 66 illustrations. 402pp. 5⅜ x 8½. (Available in U.S. only) 21252-1 Pa. $5.95

MAGIC: STAGE ILLUSIONS, SPECIAL EFFECTS AND TRICK PHOTOGRAPHY, Albert A. Hopkins, Henry R. Evans. One of the great classics; fullest, most authorative explanation of vanishing lady, levitations, scores of other great stage effects. Also small magic, automata, stunts. 446 illustrations. 556pp. 5⅜ x 8½. 23344-8 Pa. $6.95

THE SECRETS OF HOUDINI, J. C. Cannell. Classic study of Houdini's incredible magic, exposing closely-kept professional secrets and revealing, in general terms, the whole art of stage magic. 67 illustrations. 279pp. 5⅜ x 8½. 22913-0 Pa. $4.00

HOFFMANN'S MODERN MAGIC, Professor Hoffmann. One of the best, and best-known, magicians' manuals of the past century. Hundreds of tricks from card tricks and simple sleight of hand to elaborate illusions involving construction of complicated machinery. 332 illustrations. 563pp. 5⅜ x 8½. 23623-4 Pa. $6.95

THOMAS NAST'S CHRISTMAS DRAWINGS, Thomas Nast. Almost all Christmas drawings by creator of image of Santa Claus as we know it, and one of America's foremost illustrators and political cartoonists. 66 illustrations. 3 illustrations in color on covers. 96pp. 8⅜ x 11¼. 23660-9 Pa. $3.50

FRENCH COUNTRY COOKING FOR AMERICANS, Louis Diat. 500 easy-to-make, authentic provincial recipes compiled by former head chef at New York's Fitz-Carlton Hotel: onion soup, lamb stew, potato pie, more. 309pp. 5⅜ x 8½. 23665-X Pa. $3.95

SAUCES, FRENCH AND FAMOUS, Louis Diat. Complete book gives over 200 specific recipes: bechamel, Bordelaise, hollandaise, Cumberland, apricot, etc. Author was one of this century's finest chefs, originator of vichyssoise and many other dishes. Index. 156pp. 5⅜ x 8. 23663-3 Pa. $2.75

TOLL HOUSE TRIED AND TRUE RECIPES, Ruth Graves Wakefield. Authentic recipes from the famous Mass. restaurant: popovers, veal and ham loaf, Toll House baked beans, chocolate cake crumb pudding, much more. Many helpful hints. Nearly 700 recipes. Index. 376pp. 5⅜ x 8½. 23560-2 Pa. $4.95

CATALOGUE OF DOVER BOOKS

THE ANATOMY OF THE HORSE, George Stubbs. Often considered the great masterpiece of animal anatomy. Full reproduction of 1766 edition, plus prospectus; original text and modernized text. 36 plates. Introduction by Eleanor Garvey. 121pp. 11 x 14¾. 23402-9 Pa. **$8.95**

BRIDGMAN'S LIFE DRAWING, George B. Bridgman. More than 500 illustrative drawings and text teach you to abstract the body into its major masses, use light and shade, proportion; as well as specific areas of anatomy, of which Bridgman is master. 192pp. 6½ x 9¼. (Available in U.S. only) 22710-3 Pa. **$4.50**

ART NOUVEAU DESIGNS IN COLOR, Alphonse Mucha, Maurice Verneuil, Georges Auriol. Full-color reproduction of *Combinaisons ornementales* (c. 1900) by Art Nouveau masters. Floral, animal, geometric, interlacings, swashes—borders, frames, spots—all incredibly beautiful. 60 plates, hundreds of designs. 9⅜ x 8-1/16. 22885-1 Pa. **$4.50**

FULL-COLOR FLORAL DESIGNS IN THE ART NOUVEAU STYLE, E. A. Seguy. 166 motifs, on 40 plates, from *Les fleurs et leurs applications decoratives* (1902): borders, circular designs, repeats, allovers, "spots." All in authentic Art Nouveau colors. 48pp. 9⅜ x 12¼. 23439-8 Pa. **$6.00**

A DIDEROT PICTORIAL ENCYCLOPEDIA OF TRADES AND INDUSTRY, edited by Charles C. Gillispie. 485 most interesting plates from the great French Encyclopedia of the 18th century show hundreds of working figures, artifacts, process, land and cityscapes; glassmaking, papermaking, metal extraction, construction, weaving, making furniture, clothing, wigs, dozens of other activities. Plates fully explained. 920pp. 9 x 12. 22284-5, 22285-3 Clothbd., Two-vol. set **$50.00**

HANDBOOK OF EARLY ADVERTISING ART, Clarence P. Hornung. Largest collection of copyright-free early and antique advertising art ever compiled. Over 6,000 illustrations, from Franklin's time to the 1890's for special effects, novelty. Valuable source, almost inexhaustible.
Pictorial Volume. Agriculture, the zodiac, animals, autos, birds, Christmas, fire engines, flowers, trees, musical instruments, ships, games and sports, much more. Arranged by subject matter and use. 237 plates. 288pp. 9 x 12. 20122-8 Clothbd. **$15.00**

Typographical Volume. Roman and Gothic faces ranging from 10 point to 300 point, "Barnum," German and Old English faces, script, logotypes, scrolls and flourishes, 1115 ornamental initials, 67 complete alphabets, more. 310 plates. 320pp. 9 x 12. 20123-6 Clothbd. $15.00

CALLIGRAPHY (CALLIGRAPHIA LATINA), J. G. Schwandner. High point of 18th-century ornamental calligraphy. Very ornate initials, scrolls, borders, cherubs, birds, lettered examples. 172pp. 9 x 13. 20475-8 Pa. **$7.95**